FOOD AID RECONSIDERED: ASSESSING THE IMPACT ON THIRD WORLD COUNTRIES

EADI BOOK SERIES 11: SERIES ON EVALUATION OF AID
edited by Olav Stokke (This volume is also No. 73 in the series
Norwegian Foreign Policy Studies, Norwegian Institute of
International Affairs, Oslo.)

FOOD AID RECONSIDERED
Assessing the Impact on Third World Countries

edited by

EDWARD CLAY

and

OLAV STOKKE

FRANK CASS · LONDON

Published in collaboration with
The European Association of Development Research and
Training Institutes (EADI) Geneva

First published in Great Britain by
FRANK CASS & Co. LTD.
Gainsborough House, 11 Gainsborough Road
London E11 1RS, England

and in the United States of America by
FRANK CASS
c/o International Specialized Book Service Ltd.
5602 N.E. Hassalo Street
Portland, OR 97213-3640

Copyright © 1991 EADI and authors

British Library Cataloguing in Publication Data

Food aid reconsidered : assessing the impact on third world
countries. – (EADI book series, 11).
1. Developing countries. Food supply. Foreign assistance
by developed countries
I. Clay, Edward J.　II. Stokke, Olav　III. European
Association of Development Research and Training
Institutes IV. Series
363.88

ISBN 0-7146-3414-X

All rights reserved. No part of this publication may be reproduced in any form or by any means, electronic, mechanical, photocopying, recording or otherwise, without the prior permission of Frank Cass and Company Limited.

Printed in Great Britain by
Antony Rowe Ltd, Chippenham

Contents

Foreword	vii
List of Tables	x
List of Figures	xii
Glossary	xiv
1. Assessing the Performance and Economic Impact of Food Aid: The State of the Art *Edward Clay and Olav Stokke*	1
2. An Additional Resource? A Global Perspective on Food Aid Flows in Relation to Development Assistance *Ram Saran and Panos Konandreas*	37
3. The Disincentive Effect of Food Aid: A Pragmatic Approach *Simon Maxwell*	66
4. Modelling the Role of Food Imports, Food Aid and Food Security in Africa: The Case of Botswana *John Cathie*	91
5. Dairy Aid and Development: Current Trends and Long-Term Implications of the Indian Case *Martin Doornbos, Liana Gertsch and Piet Terhal*	117
6. Triangular Transactions, Local Purchases and Exchange Arrangements in Food Aid: A Provisional Review with Special Reference to Sub-Saharan Africa *Edward Clay and Charlotte Benson*	143
7. Food Aid and Structural Adjustment in Sub-Saharan Africa *Hans Singer*	180
8. An Introduction to the Sources of Data for Food Aid Analysis with Special Reference to Sub-Saharan Africa *Stéphane Jost*	191
Index	202
Notes on Contributors	208

Foreword

The present volume belongs to a series focusing on the evaluation of development assistance. This series is part of a research and publishing programme under the auspices of the Working Group on Aid Policy and Performance of the European Association of Development Research and Training Institutes (EADI). Two forthcoming volumes will focus on the evaluation policy of a selection of European and international aid agencies and on evaluation methods.

The decision to make the evaluation of development assistance its major task was reached at EADI's IV General Conference in Madrid in 1984 when the Working Group on Aid Policy and Performance responded favourably to a position paper by the Convenor. That year, the Working Group had fulfilled its first major task with the publication of *European Development Assistance*, Volumes 1 (*Policies and Performance*) and 2 (*Third World Perspectives on Policy and Performance*) as No.4 in the EADI Book Series, published by Tilburg. Three years later, at EADI's V General Conference in Amsterdam, the evaluation programme was formally launched. At that point, funding had been secured for the first project, and the planning and design of the projects could enter a new phase.

The three volumes are emerging from a similar process. For each, a design is worked out by the organisers. Outstanding researchers or experts in the relevant field are asked for comments on the design. These consultations are followed by invitations to provide a component study. The draft studies for each project are presented at a workshop to which, in addition to the authors, a selected group of experts, representing the immediate environment of the chosen theme (researchers, top administrators and others in the specific field) are personally invited. It is the matured versions of drafts which have been scrutinised through this process and then edited that are presented in this series.

The series starts with a volume on a high profile and controversial aspect of aid policy, food aid. It will be followed by one on the evaluation policies of some European countries and international aid agencies which is in the editorial phase. The third project, on evaluation methods, is being implemented.

Assessment of the role and impact of food aid has acquired considerable importance and led to many evaluations of special projects and individual countries as well as whole aid programmes. The objective set for this particular project, therefore, was to provide a stocktaking of the state of the art. The considerable body of work already undertaken offered an opportunity to digest its contribution to both the policy issues in question and the use of economics and other social sciences in impact assessment.

Contributions to this volume have been chosen to cover the wide range of issues that arise in debates on food aid but also to allow discussion of the lessons for research design and methodology that will be relevant to future research and policy analysis. They concentrate on recent experience, especially in sub-Saharan Africa, which became during the 1980s the most important recipient of food aid in quantitative terms. The large number of countries involved ensures that the region has been the focus of the greatest concentration of attention by food aid administrators and policy analysts.

The papers were presented in February 1989 at a workshop organised by the Norwegian Institute of International Affairs at Lysebu, on the outskirts of Oslo. Administrators from a number of European and international aid agencies and non-governmental organisations were also involved. This fertile interaction of practitioners and researchers resulted in a significant strengthening of almost all the papers that are now presented.

Dr Edward Clay, Director of the Relief and Development Institute, London, and co-organiser of this project, has added another dimension by linking the EADI project on food aid with the network established through a series of meetings on food aid at the Institute of Development Studies (IDS) at the University of Sussex (Brighton). Edward Clay, previously a fellow at the IDS, and Professor Hans Singer have been associated with these IDS seminars since the first in 1982. Thus, in some sense, the Lysebu seminar might be considered the sixth in the series.[1]

No research and publishing programme on this ambitious scale could proceed without much help and goodwill. The overall programme received, at an early stage, a funding commitment from the Ministry of Development Co-operation, Oslo, that made it possible to proceed with the planning. The Ministry has generously contributed to the particular project on food aid by funding the major costs involved in organising the Lysebu workshop. We acknowledge this grant with pleasure and appreciation.

From the very beginning, the programme has had the privilege of being integrated into the research programme of the Norwegian Institute of International Affairs, Oslo, and has benefited from its

FOREWORD

institutional support. The project on food aid has, in particular, benefited from the efficient secretarial support of Ms Mette Rokkum, especially during the workshop at Lysebu. The Relief and Development Institute, London, has also provided substantial institutional support for the project on food aid, and warm thanks for secretarial assistance are extended, in particular, to Ms Angela Burton. We are also very much indebted to Ms Elizabeth Everitt for her assistance in editing the present volume.

London and Oslo, November 1989

Edward Clay
Co-Organiser
Project on Food Aid

Olav Stokke
Convenor, EADI Working Group
on Aid Policy and Performance
Director, Evaluation Programme

NOTE

1. Reports of the Five Seminars are in Clay and Pryer [*1982*]; a collection of papers in *Food Policy*, Vol.11, No. 1, pp.17–46; Clay and Everitt [*1985*]; Clay and Shaw [*1987*]; and Clay and York [*1987*]. For details see references to Chapter 1.

List of Tables

Chapter 1

Table 1   Total World Cereals Food Aid Shipments and Cereals Food Aid Shipments to Sub-Saharan Africa, since 1970/71 (in Thousand Tonnes)

Table 2   Total World Cereals Food Aid by Category of Use, 1979/80–81/82 to 1985/86–87/88 (in Thousand Tonnes)

Table 3   Cereals Food Aid to Sub-Saharan Africa by Category of Use, 1979/80–81/82 to 1985/86–87/88 (in Thousand Tonnes)

Chapter 2

Table 1   Growth Rates of Total Value of Food Aid and Financial Aid and Relationship between them, 1969 to 1986

Table 2   Estimates of Real Cost to Donors of Providing Food Aid under Different Values of the Substitution Coefficient α (1971–88)

Chapter 3   None

Chapter 4

Table 1   Time Series of Exogenous Variables, Deviations from the Trend (in Percentage Terms)

Table 2   Cereal Production, Imports and Estimated Value of Food Aid, 1965 to 1986

Table 3   Short Run Policy Options in Response to Drought: Model Simulation Results for the Year 1981–82

Chapter 5

Table 1   Operation Flood II – Targets and Achievements as Recorded in Different Plan Documents

Table 2   Coverage of Veterinary Services, Artificial Insemination and Provisions for Cattle Feed under OF II (September 1985)

Table 3   Distribution of Funds under OF I and OF II

LIST OF TABLES

Table 4   Cumulative Disbursement of Funds under OF II over Various States up to mid-1986 (Million Rs.)

Chapter 6

Table 1   Total Triangular Transactions, Trilateral Exchanges and Local Purchases of Cereals, 1983/84–1987/88, Arranged by Category of Country of Purchase (in Grain Equivalent, Thousand Tonnes)

Table 2   Total Triangular Transactions, Trilateral Exchanges and Local Purchases of Cereals Received by Sub-Saharan African Countries, 1983/84 to 1987/88 (in Grain Equivalent, Thousand Tonnes)

Table 3   Commodity Composition of Triangular Transactions, Trilateral Exchanges and Local Purchases, 1983/84 to 1987/88 (in Grain Equivalent, Thousand Tonnes)

Table 4   Total Triangular Transactions, Trilateral Exchanges and Local Purchases by Source, 1983/84 to 1987/88 (in Grain Equivalent, Thousand Tonnes)

Table 5   Comparison of Triangular Transactions of Cereals Originating in Sub-Saharan African Countries with Total Cereals Exports from those Countries, 1983/84 to 1987/88 (in Grain Equivalent, Thousand Tonnes)

Table 6   Total Food Aid and Developing Country Cash Purchases and Trilateral Exchanges of Cereals (as Reported to the FAO) Arranged by Donor in Decreasing Importance of Involvement in Purchases/Exchanges, 1983/84 to 1987/88 (in Grain Equivalent, Thousand Tonnes)

Chapter 7   None

Chapter 8

Table 1   Important Food Aid Statistical Sources: Reporting Periods, Coverage, Objectives and Publications

List of Figures

Chapter 1

Figure 1   Total Cereals Food Aid Shipments and Shipments Received by Sub-Saharan Africa, 1972/73 to 1988/89

Figure 2   Index of Total Cereals, Skimmed Milk Powder and Vegetable Oil Food Aid Shipments, 1977 to 1987

Figure 3   Cereals Food Aid to Sub-Saharan Africa by Category of Use, 1979/80 to 1987/88

Figure 4   Total Cereals Food Aid by Category of Use, 1979/80 to 1987/88

Chapter 2

Figure 1   Effects of Elimination of Food Aid on World Price

Figure 2   Donor Real Costs and Increase in Consumer Expenditures for Providing Food Aid in Wheat

Chapter 3

Figure 1   Food Aid Disincentives – Summary of Scenarios

Figure 2   Food Aid Disincentives – Summary of Key Indicators

Figures
3 and 4   Sorghum Prices, Woldia, 1981–86

Chapter 4

Figure 1   Cereal Production and Food Aid 1965 to 1984 (Thousand Tonnes)

Figure 2   Resource Inflow to Botswana, 1965 to 1984 (Million Pula)

Chapter 5   None

Chapter 6

Figure 1   Per Capita Cereals Production Indices of Zimbabwe and Thailand, 1975 to 1987

LIST OF FIGURES

Figure 2   Triangular Transactions of Cereals Originating in Zimbabwe and Destined for other SSA Countries, 1986/87 (Based on FAO Data)

Chapter 7 None

Chapter 8 None

Glossary

LIST OF ACRONYMS AND ABBREVIATIONS

ADB	African Development Bank
CCC	Commodity Credit Corporation (United States)
CILSS	Comité Permanent Inter-états de Lutte contre la Secheresse dans le Sahel.
c.i.f.	carriage insurance freight
CRS	Catholic Relief Services
CSD	Committee on Surplus Disposal (FAO)
DAC	Development Assistance Committee (of the OECD)
EC	European Community
EADI	European Association of Development Research and Training Institutes
FAC	Food Aid Convention
FAO	Food and Agriculture Organisation (of the United Nations)
f.o.b.	free on board
FSG	Food Studies Group
FY	Fiscal Year
GIEWS	Global Information and Early Warning System on Food and Agriculture (FAO)
IBRD	International Bank for Reconstruction and Development (World Bank)
IDC	Indian Dairy Corporation
IEFR	International Emergency Food Reserve
ILO	International Labour Organisation (of the United Nations)
INTERFAIS	International System of Information on Food Aid (of WFP)
IRDP	Integrated Rural Development Programme
i.s.t.c.	internal shipping and transport costs
IWC	International Wheat Council
LDC	Least Developed Country
NFS	National Food Strategy (Botswana)

GLOSSARY

NCDF	National Co-Operative Dairy Federation (India)
NDDB	National Dairy Development Board (India)
NGO	Non-Governmental Organisation
NPA	Nouvelle Politique Agricole (Senegal)
ODA	Official Development Assistance
OECD	Organisation for Economic Co-Operation and Development
OF	Operation Flood (India)
PL480	Public Law 480 (United States)
PREF	Plan à Moyen Terme de Redressement Economique et Financier (Senegal)
S416	Section 416 of the Agricultural Trade Act of 1949 (United States)
SACU	Southern African Customs Union
SADCC	Southern African Development Co-Ordination Conference
SAM	Social Accounting Matrix
SSA	Sub-Saharan Africa
UMR	Usual Marketing Requirement
UNCTAD	United Nations Conference on Trade and Development
UNDP	United Nations Development Programme
UNICEF	United Nations International Children's Emergency Fund
USAID	United States Agency for International Development
USDA	United States Department of Agriculture
WFC	World Food Council
WFP	World Food Programme (of the United Nations)

1

Assessing the Performance and Economic Impact of Food Aid: The State of the Art

EDWARD CLAY and OLAV STOKKE

FOOD AID – PART OF AID POLICY

> 'When I use a word', Humpty Dumpty said in a rather scornful tone, 'it means just what I choose it to mean – neither more nor less.'
> 'The question is', said Alice, 'whether you can make words mean different things.'
> 'The question is', said Humpty Dumpty, 'which is to be master – that's all.'
>
> (Lewis Carroll, *Through the Looking Glass*, 1888)

In the now large food aid literature, the reader and the author are usually assumed to share a common definition; that is that food aid is what the donors choose to call food aid.[1] These are the aid funded food operations in a particular developing country that are characterised as food aid. More generally, food aid is that which is reported as such annually in a separate survey to the Development Assistance Committee (DAC) of the Organisation for Economic Co-operation and Development (OECD) by member countries.

Food aid certainly has an identity of its own and, to a large extent, is discussed, analysed and even administered within its own confines. Nevertheless, it is an integral part of the aid policy of donor countries. Its relative position, and indeed its substance, differ from one country to another, but so do other components of aid programmes.

The relative importance and content of food aid have long been recognised to reflect the export profile of the donor country and even its surplus of the commodity in question. For example, the USA provides a wide range of cereals and is the most important source of edible vegetable oil as aid. Canada, Japan and Norway all provide fish products as aid. Hence too, the Food Aid Convention in which cereal exporters, including the European Community as a regional grouping, have made binding minimum commitments to provide cereals aid. The FAO Committee on Surplus Disposal [*FAO, 1980*], set up to ensure that food aid did not disrupt normal export trade in foodstuffs, is another institutional arrangement linked to agricultural export trade. All these features of the food aid regime perhaps explain why the statistical reporting of food aid is distinct from, and possibly more complex than, that of other forms of development assistance.

The motives of governments for providing foreign aid are mixed; humanitarian and altruistic motives are combined with more or less explicit self-centred ones geared to satisfying economic, political and strategic interests. The emphasis differs from one government to another. Whereas the major international powers may be expected to use aid explicitly as an instrument of their foreign policy, emphasising the pursuance of economic and strategic objectives, some Western 'middle-sized' powers have emphasised humanitarian concerns and developmental objectives [*Stokke, 1989*]. Although intertwined, and with a blend that varies from one country to another, it is possible to distinguish between the following broad categories of motives for foreign aid:

- Basically altruistic, *humanitarian* concerns which are translated into two sets of objectives, namely (i) to alleviate distress and dire poverty, in its most extreme form through international relief operations, and (ii) to stimulate development, involving sustainable economic, social and even political growth of a kind that reflects values cherished in the donor country (such as human rights, democracy).
- The explicit pursuit of *self-interest* in various forms. These include economic interests related to exports, investments and the exploitation of raw materials as well as the promotion of cultural, economic or political ideologies held in high esteem by the donor government and interests related to political and strategic concerns. Here the distinction between altruism and self-interest blurs. For example, the promotion of economic liberalism may reflect perceptions of what is beneficial for the recipients of aid and at the same time be of mutual interest and involve a domestic political pay-off.

FOOD AID: THE STATE OF THE ART

How Does Food Aid Fit within these General Patterns?

Before addressing this question, the point should be made that food aid is provided in many different forms and for different purposes, and that the content of different countries' food aid programmes varies profoundly. For some countries, albeit a minority, such as Japan, the Netherlands and the United Kingdom, food aid may be considered a *relatively* integral part of the overall aid programme and governed by the norms established for aid policy. The main suppliers of food aid in kind, particularly the United States, constitute the other extreme tendency, of a programme that historically has been separate in budgeting and policy formation from other development assistance.

Much of the food aid debate has been concerned only with the policies and performances of the latter group, very often without making this important distinction. Because of the United States' preponderant role in food aid (it continues to provide about half of all commodities) the imbalance is understandable. Nevertheless, because the other donors which provide finance are relatively more important in funding international programmes and in recent non-traditional activities, such as triangular transactions, that focus can imply a distortion of the political reality.

Food as Emergency Aid

The most widely stated and publicly endorsed motive for food aid in donor countries is now humanitarian relief, providing the rationale for emergency aid through international organisations, voluntary agencies (NGOs) and directly from donor to recipient government. This applies, in particular, to food aid provided to meet needs arising from natural disasters and civil and international conflict. However, the humanitarian motive does not necessarily preclude the pursuit of self-interest as well, in the form of aid and the conditions involved.

Emergency aid has a problematic of its own. Its most vital task traditionally has been seen as meeting immediate needs arising from a catastrophe. But the actual nature of humanitarian aid and public perceptions of it, reflected in the modalities of emergency aid, have become increasingly inconsistent. Thus the greater part of emergency food aid during the 1980s went to long-term support for displaced populations and refugees, rather than to succour the victims of disasters currently the focus of media attention. This divergence of form and reality appears to be leading to the establishment of a new, non-emergency, protracted relief operations modality, specifically for aid to displaced and refugee populations.

The change will bring into sharper focus a second issue: even within the confines of relief, more long-term tasks associated with creating sustainable development have increasingly come to the fore. This also constitutes the core aim of ordinary, long-term development assistance. To what extent can food aid be considered an efficient instrument in the pursuance of this objective?

Food Aid and Long-Term Development

The crucial test for aid aimed at creating sustainable development is its instrumentality in strengthening the capability and capacity within Third World countries to improve human and material conditions. From this perspective, assistance to improve the ability to *produce* food would score higher marks than just *providing* food. This perspective has been strongly reflected in the food strategy theme promulgated initially in the late 1970s by the World Food Council (WFC) and taken up enthusiastically in the early 1980s by, for example, the EC Commissioner for Development, M. Pisani [*EC, 1981; 1982*]. However, in some situations the supply of food may be a precondition for sustainable development even in a more long-term perspective. That is the aim of project food aid, for example in food for work programmes, dairy development and nutrition projects for building human capital.[2]

From a macro-economic perspective, food aid in kind does not differ, in principle, from other types of commodity aid. It is usually tied to procurement in the donor country, often doubly and even triply tied, that is the product and even the producer are identified. It saves foreign currency (improves the balance of payments) for the recipient, provided that the government, in the absence of food aid, would have given priority to equivalent imports; or it provides additional food, provided that the government (or private trade) would not have given priority to this kind of import.

The more fundamental developmental effects of state-to-state food aid will, therefore, be highly dependent on the policy orientation and priorities of the recipient government. If the effect of food aid, in the first place, is to ease, fully or partly, the balance of payments constraints of the recipient economy, then its developmental effects will depend on how the extra foreign currency is used. The food is usually sold on the domestic market, like other forms of commodity aid. The developmental effect of local currency revenue generated will depend critically on how these funds are allocated by the Ministry of Finance. This underlies a fundamental aspect of the aid relationship.

Reaching the Poor

Most donor governments insist in their declared policies that assistance is provided to alleviate needs and create sustainable development, in particular for the poor and even the poorest. Both bilateral and multilateral aid agencies have *targeted* aid for the poor, specifically, for example, women and children. To what extent does food aid, when not directed towards relief operations, fit into such a strategy?

The answer is probably that food aid is not particularly appropriate for directly targeting the poorest, though with some outstanding exceptions (such as food for work programmes). For aid going directly into the coffers of the Ministry of Finance, the institutional setting may provide resources that are used to strengthen the country's infrastructure or formal economy rather than to meet the immediate needs of the poorest in the most remote areas. However, donors may insist on having a say in how funds obtained from the sale of food aid in the national market are used, and they often do. But this so-called monetised food aid emerges as a very cumbersome way of achieving the kind of effects aimed at – a financial transfer would have done the trick much more easily.

Combining Realist and Humanitarian Concerns

As far as development assistance motivated by *selfish political and economic interests* is concerned, a distinction has been made between aid as an explicit instrument of a realist foreign policy and aid provided within the framework of humane internationalism [*Stokke, 1989*]. In the first case, aid is exclusively intended and designed to further the security concerns or economic interest of the donor country, as perceived by its government. The explicit promotion of self-interest is also part of the motivation for aid in the second category – and indeed of the design of aid programmes; but in this case it is combined with a concern for alleviating distress, meeting basic needs and creating development opportunities in the Third World.

The balance of concerns may vary from country to country. A few governments are concerned to articulate policies consistent with an ethical perspective – the moral obligation to alleviate human suffering and meet human needs across national, political and cultural borders – but even the most idealistic tend to combine altruism with the pursuit of self-interest. The history and nature of food aid result in a complex interweaving of realist policy considerations and humanitarian concerns in all donor countries.

In the context of the African food crisis, discussed more fully below, food aid was declared, and expected to be, an expression of humane internationalism. But it also continues to be an instrument of a realist policy tradition. This is most explicitly the case in the United States, the major provider of food aid. From the inception of the PL480 programme in 1954, food aid has been instrumental in combining policy goals of altruism with narrower self interest: using food surpluses in a way that calls for approval instead of finding alternative ways to dispose of them (dumping onto domestic or international markets or even destruction) which might involve political as well as economic costs. For the donor country, food aid in kind has constituted, by definition, an integral part of a strategy to ensure a high return flow of aid, as it has been tied to procurement in the donor country.

Mobilising Domestic Support for Foreign Aid

Two additional, somewhat interrelated factors are part of the strategy to ensure a high return flow of aid. First, an important feature in the art of politics is to create alliances and thus rally different interests behind public tasks which are considered important from an overall perspective.

Food aid has attracted the support, and also the opposition, of ethical and religious coalitions with humane, internationalist goals. Like commodity aid in general, it has been instrumental in gaining the support of the producer – in this case the farmer lobby, traditionally strong in domestic politics – for foreign aid. And, as food aid in kind is often processed in, and shipped by freighters belonging to, the donor country additional domestic interest groups are satisfied as well.

The other factor relates to the balance of payments situation of the donor country – the main justification for tying aid. Food aid in kind reduces the foreign currency cost of overseas development assistance. This may affect the political support for foreign aid, as it answers the argument that foreign aid should not be financed by foreign loans as expressed *inter alia* by Scandinavian and US Conservatives. It may even affect positively the volume of aid (provide additionality). Given a domestic setting where most aid is tied, but where food prices are internationally competitive, food aid may even stand out as an effective way to use ODA grants compared with some alternative uses – as in the case of Canadian state-to-state aid.

FOOD AID: THE STATE OF THE ART

The Additionality Question

The second-best nature of food transfers explains the crucial importance of the additionality question. To what extent is food aid additional to, or substituting for, other development assistance? That question is addressed by Ram Saran and Panos Konandreas in Chapter 2. The always provisional answer is that there has been some additionality, probably, in the case of the United States. The PL480, and more recently S416, programmes have been funded from the Department of Agriculture budget and the volume of donated commodities (even the budgeted cost) has fluctuated with the availability of surpluses for export. But the Gramm–Hollings–Rudman Amendment is now shifting the attributable cost of PL480 to the Foreign Assistance budget. Even in the United States, food aid is therefore more likely during the 1990s to compete at the margin with other elements of the aid programme. This fact necessitates both a closer scrutiny of the record of food aid and a careful consideration of the options for using food aid in the foreseeable future. These issues are addressed directly by all the contributors in this volume.

Assessing the Value of Food Aid

There is a long tradition of formulating such issues as concrete questions to which quantified answers can be provided. This applies particularly to valuing a resource transfer to a recipient economy, or weighing the ultimate impact on beneficiaries of food aid actions against the costs to the donor. Precisely because it has been drawn from exportable, or otherwise unexportable, surpluses these questions have been considered more closely in the case of food aid than, for example, when applied to the extensively tied flows of services in the form of technical co-operation funded from financial aid [*Singer et al., 1987*].

Ram Saran and Panos Konandreas consider these issues once more against the background of the massive surpluses that were overhanging world cereal, edible oil and dairy product markets during the mid 1980s. The inference that they draw, not surprisingly, is that food aid transfers are of less value to recipient countries altogether, and also cost less than the value accounted to aid or agricultural budgets and reported in official statistics as food aid. But that too is only a partial answer. Would these conclusions be valid also for periods of 'tighter' international markets such as those in 1972–74 and again in 1988–89? That is an issue for further research.

There is also the issue of the distribution of the transfers and costs, both amongst and within recipient countries and amongst and within the donor economies. The narrow dividing line which Konandreas and Ram Saran explore between 'aid' and concessional sales (dumping) raises the more fundamental question of what exactly is food aid.

A Doubly Tied Commodity Transfer

Most writing on food aid is concerned with an aid transfer from economy A, let us say Australia, to economy B, let us say Bangladesh. The source country has to be country A. The commodity is a food stuff which country A agrees to provide on a concessional basis (involving concessionality of at least 25 per cent as defined by the OECD's DAC). It is an 'aid' transfer, because the transaction is mediated by an 'aid' organisation (bilateral or international) or by an NGO and, therefore, is intended to have a developmental or humanitarian objective. In practice, most of the literature on food aid has also been concerned only with cereals aid, ignoring the substantial flow of dairy products (from the USA as well as the EC), edible vegetable oils, fish products etc.

But this common sense definition of food aid has been rendered at least marginally problematical by a number of developments.

A substantial proportion of what is officially recorded as food aid, approximately eight to ten per cent of cereals aid in 1986/87 and 1987/88, was acquired by donors in developing countries through purchases with convertible currencies or through a barter or 'swap' trade arrangement. Approximately three quarters of those commodities were then targeted on sub-Saharan Africa (SSA), and half of the cereals were acquired within SSA. Perhaps a quarter of these transactions by volume in SSA also involved local purchases in the aid recipient country without any linked trade flow. This partial breaking of the link with trade has led to the suggestion that such transactions are not food aid, but aid for food. They are usually an alternative to a normal food aid action involving imports from a developing country, fulfilling donor commitments under the Food Aid Convention.[3] These developments raise operational concerns of cost-effectiveness, management efficiency and appropriateness associated with purchase and barter arrangements. In addition developmental questions arise, different from those related to conventional food aid, which concern the source economy or region in which the commodities are acquired. These developments, which are still little documented, provide the justification for including in this volume the chapter by Clay and Benson on triangular transactions.

FOOD AID: THE STATE OF THE ART

TRENDS IN FOOD AID

A simplified, but conventional account of recent trends in food aid flows would be broadly as follows. First, food aid levels declined sharply during the early 1970s in the period of the so-called 'world food crisis'. The quantity of cereals provided has never returned to the previous record levels of some 18 million tonnes in the mid 1960s, that were largely supplied under the United States PL480 Programme [*Wallerstein, 1980: Tables 3.2, 3.6; Huddleston, 1984*]. Nevertheless, the volume of food aid rapidly recovered to a level of around nine to ten million tonnes of grain or wheat equivalent in the late 1970s. Then, the coincidence of the African food crisis and a period of excessively depressed world markets in the mid 1980s resulted in annual levels of shipments moving slightly higher to around 11–12 million tonnes in 1984/85. Even higher levels were shipped in 1986/87 and 1987/88. But with a tighter market situation, and reduced emergency requirements in Africa, total cereals declined to eleven million tonnes during 1988/89, and ten million tonnes in 1989/90 (Figure 1).

The widely accepted view is that the combination of the renegotiated Food Aid Convention (FAC) in 1980, involving burden sharing within the donor community and a desire to avoid a recurrence of the negative consequences of a sharp fall away in food levels between 1972 and 1974, resulted in relatively stable, institutionally determined levels of food aid [*Parotte, 1983; Clay, 1985*]. But these levels, incidentally, met a declining proportion of the growing food import bill of low income countries.[4] In the mid 1980s the African food crisis coincided with a period of growing surplus disposal problems. Food aid levels began to rise even more through a combination of supply and demand side pressures. There was a severe upward movement in market prices, first of rice, then of wheat and maize during 1987/88. Food aid appears, for reasons discussed below, likely to fall back in 1989/90 to the levels of the early 1980s.

The Changing Face of Food Aid

This somewhat simplified account of recent trends in the volume of food aid conceals a number of difficulties from the point of view of economic analysis. First, there are the consequences of a major change in US agricultural export policy. During the 1950s and 1960s the food aid programme under PL480 was the main vehicle for commodity export management and trade promotion in relation to developing, as

FIGURE 1
TOTAL CEREALS FOOD AID SHIPMENTS AND SHIPMENTS RECEIVED BY SUB-SAHARAN AFRICA, 1972/73 TO 1988/89

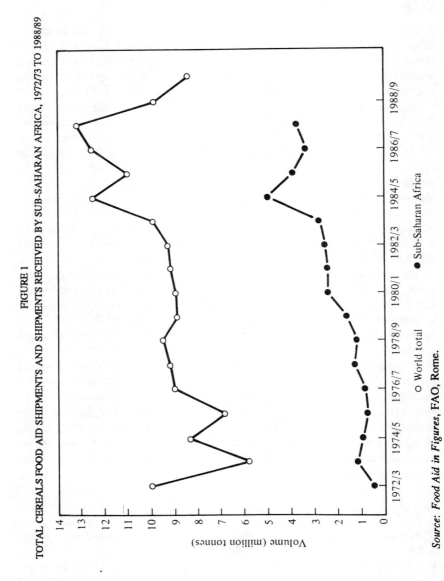

Source: *Food Aid in Figures*, FAO, Rome.

well as some developed, countries. Then, in the mid 1970s the US Department of Agriculture switched to using its own export enhancement and credit financed sales to cover a substantial part of US agricultural commodity exports to developing countries. Consequently, US food aid, then equivalent to the PL480 programme, was left as the channel for a decreasing share of agricultural exports, probably with a greater element of concessionality, but within increasingly tight administrative restrictions imposed by Congress.[5]

Second, virtually all food aid since the 1970s, apart from US PL480 Title I, has been on a grant basis.[6] Consequently, the commodity cost element has involved a higher *planned* element of concessionality. Debts may be rescheduled subsequently, but that is *ex post*, unplanned concessionality.

Third, there has been a geographical reallocation of food aid both towards least developed countries (LDCs) and towards sub-Saharan Africa (Table 1). That reallocation, particularly to Africa, has had further cost as well as programmatic implications. The concomitant shift in allocations to emergency aid, and to a lesser extent to project aid (especially outside SSA), also involved donors in meeting both a higher proportion of non-commodity costs and higher per tonne non-commodity costs (Table 2). This is in part because LDCs are given more favourable terms regarding shipping costs. Total costs of shipping commodities to landlocked destinations particularly in Africa are higher[7] and emergency assistance and project aid are also likely to involve higher administrative costs. The growing volume of acquisition in developing countries has not affected this upward cost trend since evaluations show that commodity purchases have clustered around the import parity price levels of recipient, often landlocked, countries [*Relief and Development Institute, 1987*]. The overall trend is for food aid to become more completely a full cost transfer attributable to aid budgets with growing non-commodity costs from the viewpoint of those aid agencies.

Finally, there is the question of trends in non-cereal food aid. Since the mid 1970s the European Community has been the second most important food aid donor, in terms both of physical quantities and of costs as accounted by donor agencies. For 12 years, non-cereal commodities, virtually all of them dairy products, contributed more than half the accounted costs of EC food aid [*Kennes, 1988*]. The trends in non-cereals food aid do not correlate precisely with the medium term trend in cereal transfers (Figure 2).

It seems reasonable to conclude, therefore, of food aid, as with the game of croquet in *Alice in Wonderland*, that things are not exactly

TABLE 1
TOTAL WORLD CEREALS FOOD AID SHIPMENTS AND CEREALS FOOD AID
SHIPMENTS TO SUB-SAHARAN AFRICA SINCE 1970/71 (IN THOUSAND TONNES)

YEAR	TOTAL WORLD CEREALS FOOD AID	CEREALS FOOD AID TO SSA	CEREALS FOOD AID TO SSA AS % OF WORLD TOTAL
1970/1	12,357	545	4.4
1971/2	12,513	466	3.7
1972/3	9,964	461	4.6
1973/4	5,819	1,181	20.3
1974/5	8,399	936	11.1
1975/6	6,844	747	10.9
1976/7	9,042	860	9.5
1977/8	9,211	1,282	13.9
1978/9	9,500	1,171	12.3
1979/80	8,887	1,602	18.0
1980/1	8,942	2,399	26.8
1981/2	9,140	2,402	26.3
1982/3	9,238	2,545	27.5
1983/4	9,849	2,750	27.9
1984/5	12,511	4,992	39.9
1985/6	10,949	3,879	35.4
1986/7	12,579	3,307	26.3
1987/8	13,167	3,706	28.1
1988/9‡	9,757	na	na
1989/90‡	8,400	na	na

‡ provisional

Source: FAO, *Food Aid in Figures,* various.

what they seem. The cost of food aid transfers, either in financial terms for aid budgets, or in real resource cost terms, may well have been higher during the mid 1980s than in the apparent peak period of the 1960s. Here then are some of the issues which are addressed in the component studies included in this volume.

The chapter by Ram Saran and Panos Konandreas on questions of the additionality and cost effectiveness of food transfers is especially timely. A substantial part of food aid since the mid 1970s has not been drawn from the surplus stocks of developed countries available for export, but has been financed by donors under their commitments to the Food Aid Convention, including acquisitions in developing countries. As suggested, the per tonne costs of food aid, including non-

TABLE 2
TOTAL WORLD CEREALS FOOD AID BY CATEGORY OF USE,
1979/80–81/82 TO 1985/86–87/88 (IN THOUSAND TONNES)

	1979/80 TO 1981/82		1982/83 TO 1984/85		1985/86 TO 1987/88	
PROJECT AID	6,316	24.3%	8,054	24.9%	9,024	24.6%
Agricultural & Rural Dev.	3,494	13.4%	4,001	12.4%	5,154	14.0%
Nutrition Improvement	2,293	8.8%	2,643	8.2%	3,029	8.2%
Food Security Reserves	226	.9%	794	2.5%	128	.3%
Other	303	1.2%	616	1.9%	173	1.9%
NON-PROJECT AID	15,900	61.1%	18,325	56.7%	19,331	52.6%
EMERGENCIES	3,800	14.6%	5,912	18.3%	8,394	22.8%
TOTAL	26,016	100.0%	32,291	100.0%	36,748	100.0%

Source: Based on WFP/CFA, *Review of Food Aid Policies and Programmes*, annual, various.

commodity costs attributable to the budgets of aid agencies, have probably increased.

Nevertheless, the links of food aid to surplus disposal have not been entirely broken at the level of individual commodities. This is obvious, for example, from the disposal as grants of surplus US commodities between 1983 and 1988 under Section 416 of the Agricultural Trade Act of 1949, made possible in 1982 by legislative amendment [*Bachrach, 1988*]. The large scale disposal of US skimmed milk powder stocks and other commodities as food aid under this provision has been drastically reduced with the disappearance of surplus stocks during 1987/1988.

As a result of the tighter cereal market conditions (Figure 1), there has been a decline of 25 per cent in 1988/89 cereals food aid and further reductions are anticipated in 1989/90. The major factor is that PL480 is budgeted in financial terms, so that the volume of commodities allocated is inversely related to price movements. That negative relationship has food security implications discussed below.

Aid in Dairy Products

Food aid in the form of dairy products is, if possible, even more controversial than cereals food aid. Yet these transfers have been

FIGURE 2
INDEX OF TOTAL CEREALS, SKIMMED MILK POWDER AND VEGETABLE OIL FOOD AID SHIPMENTS, 1977 TO 1987

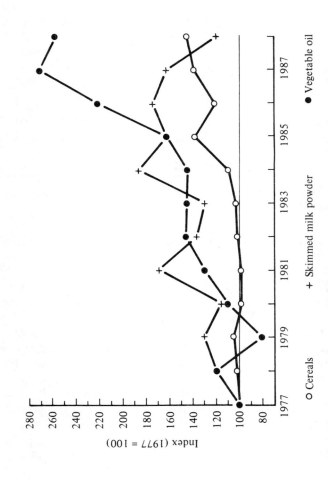

Source: *Food Aid in Figures*, FAO, Rome, various.

Note: For cereals, the figure for 1977 refers to shipments in the agricultural year 1976/77, etc.

relatively neglected in the food aid assessment literature.[8] Martin Doornbos, Liana Gertsch and Piet Terhal summarise the programme of research at the Institute of Social Studies (ISS), The Hague, which probably represents the most substantial attempt to fill the gap in our knowledge. In particular they address the difficult tertiary level research problem in reviewing the considerable and conflicting literature on Operation Flood in India.

The ISS review shows that the massive evaluative and research interest in Operation Flood has not ended controversy surrounding the programme. This is in part because views of protagonists are only partially based on the record of performance and impact. Nevertheless, a careful sifting of the evidence does narrow the range of empirically sustainable interpretations of the record. There has been no dramatic nutritional impact, either negative, in the producing communities, or positive in the consuming urban areas. The additional incomes from milk production have been distributed inequitably, following a pattern similar to the experience of other major Indian rural development programmes. There is also great regional variation in the economic impact and how the institutions perform. Nor can the increases in production that have occurred be attributed more than partially to what has been a very expensive development programme.

The increase in milk production achieved by the so-called white revolution seems to have been brought about by a combination of factors. Retail price controls have sustained the growth in demand, but largely among middle and upper income groups in urban areas. Producers have also been subsidised in the marketing, processing, credit chain. Economists have pointed to the possibilities of using funds generated by monetising food aid to achieve an incentive set of prices for producers whilst not excluding consumer subsidies. But the considerable research and extension effort to increase milk yields through cross-bred animals has been a failure. Unlike the Green Revolution innovations, technical change has played a minimal role in the growth process. None of the evaluations of Operation Flood undertaken from a donor perspective has been able to conclude unequivocally that the benefits have justified the considerable transfer of resources financed by European taxpayers through the Common Agricultural Policy and the European Community Aid Programme.

IMPACT OF FOOD AID ON AGRICULTURAL DEVELOPMENT

A combination of reasons make it particularly worthwhile at this moment to look again at the relationships between food aid, agricultural development and the provision of food security in recipient countries. In reviewing experience since the significant changes in the institutional arrangements for food aid in the mid 1970s three issues stand out.

First, there is the long-controversial issue of the positive or negative impact of food aid on agricultural development (and development more generally) in recipient economies.

Second, there is the issue of the food aid response to the African food crisis, the dominating set of events in food aid policy of the 1980s. The drought was ultimately perceived as a grave threat to the survival of large populations and to sustained economic development almost at a continental level. Did the nature and timing of the apparently massive donor response significantly ameliorate the effects of the natural disaster on affected peoples and economies? Did that assistance aid or hamper the subsequent recovery from drought?

Third, in the light of the continuing, substantial short-term variability in world food markets, there is the question of whether food aid overall has become, as was envisaged in 1974, an instrument promoting food security in recipient economies. For that to have occurred, the flow of food aid resources would have to have become counter-cyclical, counteracting the impact of international market variability on national food supply and, indirectly, on the sources of short-run, transitory food insecurity affecting vulnerable groups in recipient countries.

We shall address the issue of the impact of food aid on agricultural development first and then return to the two other issues in subsequent sections.

Food Aid and Agricultural Production – Perspectives up to the Early 1980s

A number of analysts, in reviewing the published evidence up to the early 1980s on the impact of food aid, arrived at broadly similar conclusions [*Clay and Singer, 1985; Isenman and Singer, 1977; Maxwell, 1982; Mellor, 1987; Stevens, 1979*]. The debate on the effect on a recipient country's economy, in particular its agriculture, was inconclusive, hence the continuing controversy. Important issues isolated in economic analysis were:

– the extent to which the food transfer provided foreign exchange savings or budgetary support (in the former case there was only substitution of concessional for commercial imports).

- the direct impact of imports on food production and consumption, including the pattern of distribution of these effects.
- the role of other associated agricultural policy measures.

The difficulty of the analysis was that these potential effects could not be accurately assessed in isolation, most obviously because of interaction between production and consumption, and also because of the dynamics through time of general economic activity and the evolving forward and backward linkages of the food system [*Clay and Singer, 1985: 27*].

Where quantitative economic analysis had been attempted, this indicated that the overall economic and agricultural sector impact had varied considerably between countries. That impact in turn had been heavily determined by the specifics of national food policy: the segmentation of markets, separation of consumer and producer prices, as well as overall investment policy. In making that assessment of the evidence, problems were identified that were seen to justify continuing research, particularly using formal general equilibrium models or possibly simulation analysis:

> The review of the Indian and Colombian literature ... underscores the importance of analysing the macro impact of food aid within a formal general equilibrium framework. Even in a qualitative sense, the impact remains difficult to establish whilst the analysis is undertaken in one or more of the following ways:
> (a) a partial equilibrium analysis of the agricultural sector, and especially where the analysis focuses only on aggregate food supply; or,
> (b) a wider informal general equilibrium analysis as a series of partial analyses of various potential effects;
> ... the great weakness of an informal analysis is that, where effects run in contrary directions, the result is necessarily inconclusive. This methodological inadequacy of partial analyses (even where undertaken within a formal general equilibrium framework) perhaps tempts both the critic and defender of food aid into straining the analysis to sustain an unequivocal conclusion. [*Clay, 1983: 162–3*].

In drawing that conclusion, there was an expectation that formal quantitative modelling would offer the opportunity to explore the role of food imports (supported by food aid) in a wider, macro-economic context and so move the debate forward.[9]

A further limitation of the earlier literature is that it focused on a small number of Asian and Latin American countries that had been long-term recipients of cereals food aid. Empirically, there is a need, therefore, to look at the potentially different experience in sub-Saharan Africa as well as those other developing countries that have continued to be significant recipients of food aid such as Bangladesh, Sri Lanka and Egypt.

The African Experience 1982–86

There is now an opportunity to look at what has been learned from the analysis of more recent experience, particularly in SSA. The two contributions to this volume by Simon Maxwell and John Cathie illustrate respectively the potential for using the so-called informal general equilibrium approach and formal general equilibrium modelling of African data. These authors suggest that both approaches have a place, not least because the limitations of data and the costs of general equilibrium modelling will limit the number of cases in which the latter approach can be applied.

There are also a number of policy reasons for looking again at the issues of disincentives and the impact on food consumption and the economy more widely. For more than a decade, the declared objective of food donors has been to make food aid an instrument to promote, rather than hamper, agricultural development. In particular, the intention has been to contribute to increased food security by integrating food aid into a national food strategy in recipient countries, particularly in Africa. The Maxwell and Cathie contributions demonstrate the capacity of formal economic analysis at least to indicate a plausible range of outcomes for the impact of food aid to set against these laudable objectives.

There is a further matter that requires elucidation. As suggested above, the question of the economic impact of food aid now needs to be located in a discussion of the overall implications of food imports by developing countries. The greater part of cereal imports is not provided as food aid. And a substantial part of these commercial or non food aid imports have been financed under USA soft credit programmes or EC export subsidisation programmes. An assessment of the impact of developed country export programmes comparable to the early studies of the impact of PL480 in the case of India up to 1970/71, ought now to take into account export enhancement programmes.[10]

FOOD AID: THE STATE OF THE ART

SOME LESSONS EMERGING FROM THE MANAGEMENT OF THE AFRICAN FOOD CRISIS: 1982–86

The food crisis of the 1980s in Africa has now been extensively documented. The donor response to these events has led to a large number of evaluations. There is, in addition, a substantial body of independent reporting and a limited number of economic analyses, mostly for individual countries. Most of the salient features of the food aid response to that crisis, and subsequent learning by donors, and indeed governments and NGOs, have become clear.

The crisis, moreover, spurred on the improvement in the supply of data on food aid during the 1980s because donors needed access to better information when making their operational decisions. The chapter by Stéphane Jost provides a brief and useful summary of available sources of data on food aid both globally and with special reference to sub-Saharan Africa. These are the data which most of the authors contributing to this volume, and other researchers, are using in analysing food aid issues.

A Test Case for the Food Aid System

The African food crisis has been the most important test to date of the credibility of the food aid system as reconstructed after the world food crisis of 1972–74.[11] A large number of countries moved almost simultaneously into increasing staple food deficits. The crisis began with the onset of drought in the Sahel. Subsequently, the drought became more widespread, including many countries of east and southern Africa in 1983/84. The severity of the drought was also aggravated by the enfeebled condition of many poorly managed economies severely hit by the second oil price hike. There was growing recognition from 1982 onwards that agriculture in many African countries was moving towards a crisis point. Nevertheless, the monitoring of events and articulation of a coherent strategy for responding to that crisis as it came to involve a large number of countries, was both tardy and initially unsatisfactory.

Part of the difficulty in devising a response was the large number of countries with relatively small populations and aggregate food requirements involved, compared with traditional food aid recipients in Asia. The poorly articulated, badly integrated and indeed disintegrating marketing and distribution systems exacerbated the technical response problem. Because the crisis was initially presented as being drought-related and continent-wide, international attention was probably diverted from the graver conditions rapidly developing in a few countries, particularly Ethiopia, Sudan, Mali, Chad, and subsequently in Mozambique. Early warning systems were ineffective and the organisation of a response at national and international level was too small and too late in a number of cases. There was inadequate monitoring of agricultural problems and the nature and extent of stress on vulnerable populations in several countries. There was also, from a humanitarian perspective, criminal failure of responsibility on the part of some governments. Amongst donors, there was excessive risk aversion characterised by cautious incremental responses until, finally, the whole donor community was galvanised into a massive crisis management exercise by media coverage of famine conditions in Ethiopia.

On the positive side, many countries coped extremely well with the effects of drought on food supply and entitlements through combinations of commercial and food-aided cereal imports, increased public distribution and relief operations, and management of markets. Botswana, Kenya and Zimbabwe are examples in east and southern Africa, and some of the partially affected coastal and landlocked Sahelian countries also coped relatively well. To put the international response in perspective, the relief operations were financed, in order of significance, by the domestic resources of affected countries, by aid donors, and last, and very much least, by NGOs and media-centred fundraising.

Once the situation was seen as quantitatively different, a continent-wide crisis, there was clear additionality of response (Figure 3). During the 1980s, emergency and developmental programmes appear to have been competitive, alternative, outlets for cereals food aid within a largely institutionally determined total. The only clear exception was 1984/85 when both emergency and programme assistance to Africa, and globally, increased simultaneously. The evaluative literature indicates that the massive relief operations were successful in saving lives and limiting distress and economic disintegration in Ethiopia and other affected countries. The scale of food aid operations was such that, for many countries, food aid became temporarily a large part of overall development assistance (30 per cent of aid from OECD members to the most affected countries in 1984).

On the negative side, the lagged response to the crisis resulted in much of the additional emergency food arriving after many lives had already been lost and, in some instances, when it was no longer required. With favourable rains, production rose in 1985 and 1986 to

FIGURE 3
CEREALS FOOD AID TO SUB-SAHARAN AFRICA BY CATEGORY OF USE, 1979/80 TO 1987/88

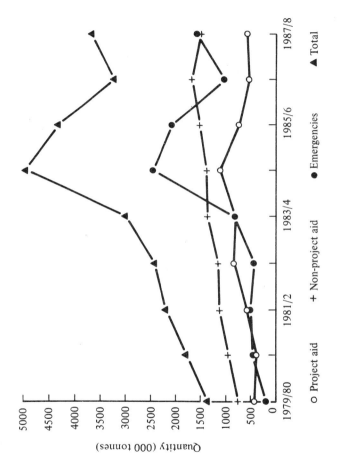

Source: WFP/CFA, *Review of Food Aid Policies and Programmes*, annual, various.

near record levels year on year in many of the most severely affected countries, particularly in the Sahel. There is some evidence of possible disincentive effects in Sahelian countries and the Sudan and, as Maxwell hints in his chapter, in Ethiopia. Indisputably, the value of tardy assistance in terms of saving lives and livelihoods, thus preventing an erosion of human capital, was diminished.

The severity of logistical problems pushed donors and governments closer together in co-ordinating food aid programming and delivery through information sharing and attempts at consistent scheduling of shipments at regional and national level. At least temporarily, effective local liaison groups emerged, involving governments and donors, that could have played a valuable continuing role in effective food aid planning after the crisis. However, in the event, donors and governments appear to have lapsed in some cases into more bilateral, less co-ordinated relationships.[12] Information sharing at an international level on food aid to so many countries was unprecedented, but encouragingly indicative of agencies being able to come to terms with more complex multi-donor food aid systems.

The Role Played by NGOs

The NGOs provide very little food aid from their own resources. However, before the African emergency, CARE and Catholic Relief Services (CRS) in particular had played an important role in US project and emergency food aid under PL480 Title II. Then, during the African crisis, the donor agencies began to use NGOs both to channel aid to recipient countries and to deliver assistance to the affected populations on an unprecedented scale. NGOs appear to have continued as a consequence to play a relatively large part in the organisation of food aid to SSA, accounting for 21 per cent of cereals and 30 per cent of other food aid to that region in 1988, a higher proportion than to any other region.[13] In Europe, Euronaid, for example, was established as a conduit for EC food aid channelled through European and international NGOs. Previously, NGOs had only been significant as handlers of dairy aid. Since the African crisis they have continued to play important and unorthodox roles, for example in cross-border operations (as in Tigre and Eritrea) and in areas where international and bilateral agencies are not able to operate directly (southern Sudan). Provisionally, they appear to have assumed a larger role in a more closely managed food aid system during the 1980s, and they represent an alternative channel to multilateral or direct bilateral assistance. The implications of this development, if it is sustained, are at this point a matter for speculation. For example, an insistence on the use of NGOs as the channel for food aid could be seen as a new form of donor conditionality.

The Main Lessons

As with the earlier crisis of 1972–74, a traumatic set of events appears to have been internalised as lessons for administrative practice within donor agencies and recipient country governments as well as NGOs.

First, the crisis generated considerable interest in early warning systems as, in effect, a trigger to emergency food aid. Resources for such systems have been maintained at a high level, even if some of the plethora of voluntary initiatives has faded away. Second, food aid donors re-examined their emergency procedures and changed regulations so as to streamline responses. Third, donors also appear likely to respond more rapidly to appeals for international emergency assistance in order to avert the costs in terms of human distress, as well as the political fall-out, of a repetition of the Ethiopian and Sudanese famines of 1984–85. Emergency assistance, at least temporarily, has come to have a relatively higher profile and priority within the food aid regime.

A problematic consequence of this change of priority is that, as noted above, there appears to be substitution between emergency and programme food aid, probably reflecting the priority accorded to the fluctuating scale of requests for provisional emergency aid (Figure 4). If developmental programme food aid has, in effect, become a residual category, this would appear to restrict the potential for a positive developmental impact. It would limit the scope for multi-annual programming, linking monetised resources to particular development activities as envisaged in most proposals for food aid reform during the past decade. At worst, there is a danger that some bilateral food aid programmes have been recognised as all too useful as ministerial slush funds for responding publicly to disasters the world over.

The lagged responses to the African crisis have drawn attention to the slow and cumbersome procedures of many donors. The programming decision process is slow. The mobilisation of commodities, and the organisation of processing and delivery, result in a total response time from request to delivery, even of so-called emergency assistance, of several months and in many cases more than a year [*Borton, 1989; US General Accounting Office, 1986; Cornelius, 1987*]. Such inflexibility limits the role that food aid, provided in a reactive mode, can play to ameliorate problems of food insecurity and substantial, unanticipated variability in food supply. Recognition of this

FIGURE 4
TOTAL CEREALS FOOD AID BY CATEGORY OF USE, 1979/80 TO 1987/88

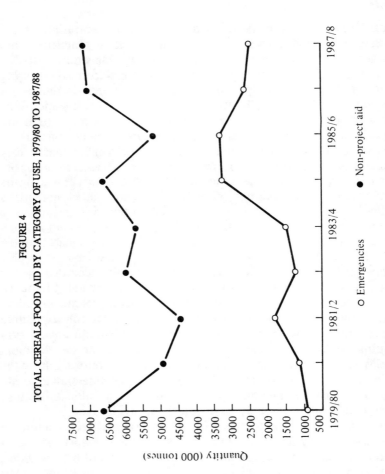

● Non-project aid ○ Emergencies

Source: Based on WFP/CFP, Review of Food Aid Policies and Programmes, annual, various.

inflexibility has also stimulated a variety of attempts to make food aid more responsive to a varying situation at regional, national and international levels.

There have been attempts to use food aid to build up directly or to finance food security or emergency stocks in individual recipient countries. Despite patchy results, such reserves at a national or regional level, financed by donors, continue to find favour in recipient countries. But donors and international funding organisations are concerned about food system management issues as well as costs.[14]

The pre-positioning of relatively small stocks at convenient shipment points for rapid emergency responses could obviously reduce the lead times for emergency actions when they have been approved. Apart from very small WFP stockpiles in Rotterdam and Singapore, little has been attempted and there has been a loss of interest on the part of donors in establishing such stocks.[15] Borrowing food from development programmes, re-routing ships, and so forth are seen as more practical, lower-cost, responses by some. Nevertheless, such responses again imply a willingness to sacrifice developmental objectives to meet immediate overwhelming, humanitarian priorities.[16]

'Concertina Projects'

Food for work or nutritional supplementation projects that can rapidly expand or contract according to the food and economic situation, so-called 'concertina projects', have played an important role in providing a food security net, particularly in South Asia. The most notable example is Bangladesh, where the expansion of food for work and the Vulnerable Group Feeding programmes has been used for post-disaster employment creation and income supplementation on a vast scale. The sheer volume of commodities involved also makes these operations a factor in cereal market management [*Hossain, 1987; Relf, 1987*].

Project design and management in a number of African countries where agricultural production is highly sensitive to drought appear to have been evolving in a similar direction. The expansion of food for work was mooted as a response in Lesotho after drought in 1983, but there was concern about the use of development projects in a relief mode [*Bryson, 1986*]. In Botswana, expansion of the nutritional supplementation programme was used for nationwide rural drought relief during the extended drought in the 1980s [*Borton, 1986*].

The concertina project concept makes it possible to use an agricultural or human resource development project in a relief context.

The provision of relief to affected people in their own location has positive developmental implications preventing socio-economic disintegration through the displacement of people and disposal of assets, as well as ameliorating the effects of food insecurity. However, experience in Asia, for example Bangladesh, as well as in Africa, raises doubts concerning trade-offs between these gains and the effectiveness of food for work activities in creating rural infrastructure and income generating assets. The quality of labour-intensive investment is likely to suffer. What can be undertaken in a crisis on a larger scale is less well designed and managed. Development projects may also acquire a relief, make-work character through use as an emergency programme [*Clay, 1986*].

Commodity Acquisition

Recent experience in Africa appears to have increased donor awareness of the need for food aid commodities to be appropriate and for more flexibility in their acquisition. The inappropriateness of distributing wheat and rice in coarse grain and tuber crop rural economies has stimulated the triangular local purchase and swap operations discussed above. Such flexibility in financing acquisition from non conventional sources increased considerably when temporary surpluses appeared as public and private stocks in many African countries in 1985/86. A number of donor agencies have modified their procedures to make at least limited commodity acquisition in developing countries mandatory. As suggested in the chapter by Benson and Clay, such practice is likely to become more commonplace in food aid operations.

One final doubt must remain concerning the food aid response to the African emergency. The crisis occurred in a large number of countries whose aggregate staple food consumption is relatively modest compared with the large, densely peopled countries of Asia. There was only an increase of 20 per cent in cereals food aid at a time when world markets were overhung by large stocks and exporters were competing with increasing intensity for markets. The response to that crisis, therefore, is little indication of how the food aid system would have coped with a worse food crisis in Asia that might plausibly have been linked to a rapid upward movement in world market prices. The continuing importance of ensuring international food security has been illustrated by subsequent events.

FOOD AID AND INTERNATIONAL FOOD SECURITY

Overall, food aid allocations by donors have been closely related to their minimum contribution levels under the FAC [*Clay, 1985*]. Nevertheless, total food aid levels, because of the behaviour of a few large donors, have remained sensitive to price movements and the relative tightness of markets for most commodities.

In the United States, the PL480 programme is budgeted on a fiscal year basis in financial terms. As a result, the actual physical allocations and shipments have been sensitive to short-term price movements – downwards in fiscal year (FY) 1979 and again in FYs 1989–90. The shipments under Section 416 are directly related to availability of surpluses for export. The supply of individual commodities such as rice or vegetable oil under PL480 has also been determined on a year-by-year basis, in terms of surplus availabilities for export. However, in accord with the Gramm–Hollings–Rudman Amendment, intended to reduce progressively the US federal deficit, there are now other pressures for reduced financial commitments accentuating the consequences of higher cereal prices on total commodity availability [*World Food Council, 1989*]. An illustration of the knock-on effects that result from a sizeable reduction in the overall programme was the substantial year-on-year cut in the United States pledges to the International Emergency Food Reserve (IEFR) during 1989 [*World Food Programme, 1989a*].

A close analysis of the short-term changes in allocations by some other food aid donors also indicates that allocations are sensitive to prices and the level of national stocks available for export. The pattern of Canadian allocations parallels very closely that of the USA during the last decade. In Australia, when drought reduced wheat production severely, there was a switch to rice which, because of the equivalence established under the FAC, resulted in a substantial reduction in actual shipped tonnage of cereals.[17] Allocations of dairy aid, although falling already as a result of political decisions to reduce the EC programme in the mid 1980s [*Kennes, 1988*], have fallen even more sharply with rising prices and the disappearance of bulk US and EC surplus stocks in 1988 and 1989 (Figure 2).

The overall implication is that food aid is still not a counter-cyclical element in the world food economy, that might buffer the effects of market movements on recipient countries. Overall, the aggregated consequence of donor decisions is to leave the level of food aid allocations and shipments still significantly pro-cyclical. Bearing in mind that the greater part of food aid is now targeted onto least-developed/low-income countries that is a serious weakness of the existing international arrangements and reflects negatively on the policies of some larger donors. In retrospect, it now appears only fortuitous that the African food crisis coincided with a period of overhanging surpluses. The response might have been more limited and possibly even more tardy if there had been a tighter market situation than the one that prevailed from 1983 to 1986.

Limitations of Food Aid as a Security Mechanism

The FAC commitments have resulted in cereal food aid levels fluctuating at around 90–120 per cent of the 1974 World Food Conference target of ten million tonnes, but with a still clearly pro-cyclical pattern of fluctuations, though much less severe than experienced from 1972 to 1976. The partial success of international agreement in making food aid an instrument of international food security, on however modest a scale, ought to be considered in the light of recent developments and possible scenarios for the 1990s.

Consider, for example, the implications of the 1987/88 drought in India having continued during 1988/89. The scale of available food aid would have been insufficient to make significant contributions to drought mitigation in India. The Indian government would have had to rely on large-scale commercial or credit purchases in a tighter market. That scenario was, on the basis of past climatological patterns of occasional two-year periods of widespread drought, a serious possibility.

Another possibility is that a severe, unanticipated, reduction in the harvest in food-aid-dependent Bangladesh could have generated additional import requirements beyond levels at which existing global food aid commitments could have made any significant impact. If Bangladesh, which received 1.6 million tonnes of cereal aid during 1988/89 and imported a total of 2.2 million tonnes of cereals, were stricken with a drought rather than a flood, import requirements could plausibly be pushed up to 3.5–4.0 million tonnes and above. If food aid were only able to cover perhaps 1.5–2.0 million tonnes of the deficit, a plausible assumption if overall levels fall to the 9.0–9.5 million tonnes level of the early 1980s, the residual import requirements would strain seriously that country's capacity to finance purchases on a commercial basis. Such quantities of commercial imports, costing $400–500 million, would also crowd out other imports necessary to sustain economic growth in one of the poorest economies, with a population exceeding 100 million and expanding at around 2.5 per cent a year.

There are two implications of such scenarios. First, food aid can no longer be considered a significant instrument of food security for countries such as India that might need to import cereals in quantities that are non-marginal in relation to world trade. Second, the transitory food security problems of countries that are in turn non-marginal in relation to overall food aid levels, such as Bangladesh, cannot be assured through food aid. The corollary is that food aid is in fact only to be considered a significant instrument of food security at a national level, or regionally, for countries whose potential import requirement is marginal to world cereal trade and to internationally agreed minimum food aid levels. In that sense, it is not merely that sub-Saharan Africa has experienced the most severe short-term transitory food security problems, but that the scale of resources required has allowed emergency food aid (an increased allocation of other categories of food aid as discussed below) to be seen by donors as potentially the most appropriate and convenient response.

These experiences and possibilities justify looking yet again at other compensatory financial mechanisms for providing a buffer against the economic effects of variability in domestic food production and world market prices on food import bills. The Compensatory Financing Facility of the International Monetary Fund (IMF), which provides for abnormal cereal import costs, was a first step in that direction. The increasing role of programme lending by development banks and bilateral donors is another area of growing flexibility. But the process is still ponderous. Perhaps when the economic costs, especially for low income countries, of cereal price variability in the late 1980s have been assessed, it will provide fresh impetus for international initiatives in this area.

FUTURE FOOD AID POLICY

This introductory chapter has so far concentrated on questions of the performance and impact of food aid, albeit focused on the more recent past. If it may be concluded, even tentatively, that the interaction of both donor and recipient policy is important, even decisive, in determining the impact of food aid, then it is appropriate to consider some of the implications of what is being learned from the record for future policy. The agricultural sector, dairy development, agricultural trade and the use of commodities from within developing countries are important areas of that discussion.

So, also, are the problems of access to accurate data on food aid, identified in the chapter by Stéphane Jost, who also makes suggestions for rationalising the system of data collection. Some of the improved range of data which became available during the 1980s, for example FAO's Global Information and Early Warning System (GIEWS) on food aid needs and WFP's INTERFAIS database, which is intended to record every food aid shipment, are not generally accessible to most academic researchers. Only processed data in published form, such as FAO's widely circulated *Food Aid in Figures*, have a general circulation. Yet technical advances are providing ways to resolve some problems, such as those of incompatible data, which have proved a limitation in the past. For example, the INTERFAIS database will be able to supply data for any predetermined period and for different points in the stages of the food aid process. As more information is held in electronic databases, perhaps conventions and initiatives are required to facilitate access to and use of these by independent researchers.

There will also be changes in the context of food aid policy. Few analysts in the 1970s anticipated the shift in resources towards Africa and the increasing importance of emergency assistance, invalidating much of the early discussion. At this time, few policy analysts appear to envisage any massive increase in food aid in the foreseeable future. Currently, perhaps influenced by short-term supply pessimism, most expect contraction or consolidation. More importantly, the closer monitoring of experience during the past decade by donor agencies has fostered a tendency for food aid to become an increasingly closely managed, but smaller, element of development assistance. There would appear to be scope for concentrating that limited resource on a number of practicable humanitarian and developmental goals, such as supporting structural adjustment in least developed African countries.

The redirection of food aid towards Africa (Table 1) was, to a considerable extent, reactive, as reflected in the high proportion of it being emergency assistance (Table 3). That emergency assistance has included short-term responses to the acute *de facto* effects of natural and man-made disasters as well as continuing assistance to countries with chronic food problems. A currently dominant theme in discussions of national and sectoral policy is that of structural adjustment. Professor Singer in his, as always, constructive and provocative manner, looks at the opportunities for food aid to play a potentially positive role in such a process, especially to ameliorate the human costs of restructuring.[18] The rehabilitation of a number of severely weakened economies in SSA requires levels of short-term balance of payments support that could be met partially by food aid. There is also a

TABLE 3
CEREALS FOOD AID TO SUB-SAHARAN AFRICA BY CATEGORY OF USE,
1979/80–81/82 TO 1985/86–87/88 (IN THOUSAND TONNES AND PERCENTAGES)

	1979/80 TO 1981/82		1982/83 TO 1984/85		1985/86 TO 1987/88	
PROJECT AID	1,368	25.5%	2,772	26.6%	1,825	16.2%
Agricultural & Rural Dev.	655	12.2%	1,409	13.5%	902	8.0%
Nutrition Improvement	428	8.0%	592	5.7%	502	4.5%
Food Security Reserves	184	3.4%	461	4.4%	111	1.0%
Other	102	1.9%	310	3.0%	316	2.8%
NON-PROJECT AID	2,849	53.1%	3,903	37.5%	4,725	42.1%
EMERGENCIES	1,146	21.4%	3,730	35.8%	4,685	41.7%
TOTAL	5,363	100.0%	10,404	100.0%	11,236	100.0%

Source: Based on CFA, *Review of Food Aid Policies and Programmes*, annual, various.

potentially significant role for food aid in supporting the process of food sector restructuring.

Financial Assistance: Aid for Food

The potential demand for food imports by Third World countries is likely to continue to grow rapidly for the foreseeable future, and with serious consequences if it is unsatisfied due to foreign exchange limitations. First, an inability to finance food imports could, as noted, constrain the development process. Secondly, problems of extensive malnutrition cannot be addressed without expanding aggregate food supply more rapidly than the foreign exchange constraints of many least developed economies will permit. Some analysts also see considerable complementary opportunities for using increased food aid to finance the development process [*Singer, 1987*], and also directly in labour-intensive rural investment, even in Africa [*Mellor, 1987*].

The first concern is to increase access to financial resources for importing food as well as other raw materials and intermediate products. Conventional food aid may not be cost-effective (Ram Saran and Konandreas) or be a source of appropriate products on a sufficiently large and predictable scale.

There are severe practical difficulties awaiting the substantially more ambitious, direct developmental uses of food aid. The recent record in many least developed countries, particularly in Africa during and after the food crisis, has underscored the difficulties of programming aid-financed imports to economies with highly variable import requirements. In contrast with the densely settled economies of South Asia, there are few SSA countries where the potential scale of labour availability or organisational capacity appears to favour large-scale rural works or other direct uses of food aid commodities on a continuous basis [*Clay, 1986*]. How is labour-intensive development to be financed in years when food imports are not needed? In addition, as already noted, conventional food aid has been unable to provide appropriate cereals on any scale for many coarse grain and tuber food economies.

To urge an increased role for food aid in financing the development process implies that food is less costly than financial aid. Even if the juggling of agricultural and aid budgets makes that so for some donors, the economic resource costs for developed countries are now probably not lower, and may even be higher, than financing food imports or supporting rural development directly.

If food aid is provided in increasingly flexible ways, with purchases in developing countries, through monetisation rather than direct distribution, the distinction between food aid and aid for food, or financial assistance, will become less important. That is potentially how much of the conventionally doubly-tied food aid will gradually disappear during the next decade.

There are of course other aspects of food aid that are given relatively little attention in this volume. These include direct, project uses for nutritional and income transfer purposes, food for work, and other agricultural and rural development activities. Emergency aid, which was the focus of two of the IDS food aid seminars [*Clay and Everitt, 1985*; and *Clay and York, 1987*] is also not directly considered in a separate chapter. Nevertheless, many of the issues that are raised in fact directly involve all food aid: issues of disincentives, and the implications of acquiring commodities in developing countries are but two examples.

It is hoped that these introductory remarks will set the context in which the contributions to this volume can be considered. In turn we hope that the volume will stimulate further lively and constructive discussion. There are no simple answers to the questions that have been posed about food aid. That is why the controversy continues.

FOOD AID: THE STATE OF THE ART

NOTES

1. See, for example, Clay and Singer [*1985*]; Clay [*1983*]; Clay [*1985*].
2. The recent literature on food for work is reviewed by Clay [*1986*]; dairy development is discussed in the chapter by Doornbos and others in this volume. Hay and Clay [*1986*] discuss the possibilities of using food for work for the creation of human resources (capital). Chapter 7 of Clay and Singer [*1985*] includes a review of the literature.
3. For an account of the evolution of the Food Aid Convention up to 1980 when a minimum contribution of 7.6 million tonnes was agreed by food aid donors, see Parotte [*1983*]. Clay [*1985*] reviews the behaviour of individual donors in meeting those minimum contributions.
4. See, for example, the various papers presented at the Ad Hoc Panel Meeting on Food Aid Projections for the Decade to the 1990s, 6–7 October 1988, National Research Council, Washington, DC.
5. For example, the share of concessional sales under the Title I Programme, mandated to low income countries as defined in terms of Gross National Product (GNP) by the World Bank, was set at 75 per cent. A fixed minimum level of commodities under the grant Title II Programme was established. A minimum mandated share of commodities was to be provided under the *de facto* grant Title III component of the Title I Programme. More recently, minimum targets for monetisation of grant food aid were also set. An example of intended increased additionality was the Title III Programme initiated in 1979 allowing, in the unhappy language of the legislation, loan forgiveness where the recipient satisfied policy conditions in the agreement.
6. The other exception was Japanese aid as so-called Kennedy Round Credits up to the early 1980s, provided to Asian countries including South Korea as well as Bangladesh.
7. Emergency and project assistance is invariably on a c.i.f. basis. In the case of such assistance to landlocked countries, food is normally provided *rendu destination*, that is, up to the border. In low income countries some donors are also meeting part of internal shipping and transport costs (i.s.t.c.).
8. For a brief review of the literature, see Clay [*forthcoming*].
9. The cumulative effect of that Indian research effort had been to narrow considerably the range of probable outcomes. Sectoral modelling suggested that the cumulative effect on agriculture ranged from a possible 15 to 30 per cent *reduction in food grain* output to an overall *increase in agricultural production* with *higher* economic growth rates and food consumption levels. Nevertheless, as Blandford and Plocki [*1977*] had shown in their careful review of Indian sectoral models, the conclusions to be drawn from such exercises are likely to be highly sensitive to choices in model specification. [*Clay, 1983: 164*].
10. In 1987, export subsidies on wheat and flour by the USA and the EC countries amounted to about $1.6 billion, or more than half the cost of food aid reported by the OECD member countries, $3 billion.
11. This account of the African food crisis draws heavily on Borton and Clay [*1986*].
12. For example in Tanzania [*Clay and Benson, 1988*] and also in Kenya (unpublished interviews by E.J. Clay with staff of donor agencies in Nairobi, January 1986).
13. 1988 is so far the only year for which there is relatively complete data from WFP's INTERFAIS database on the role of NGOs in food aid. From the USA some 15 per cent of cereals and 21 per cent of non-cereals were channelled through NGOs compared with 12 per cent of all cereals and 17 per cent of all non-cereals food aid in that year [*WFP, 1989b*].
14. World Food Programme [*1985*] evaluates earlier disappointing experiences with donor finance for emergency food.
15. For example, proposals to preposition were floated at the FAO Committee on Food

Security in 1986, but stalled for lack of donor interest. Apart from concern about stocks, there is continuing scepticism about the feasibility of being able to use prepositioned stocks. These might have to be drawn down and re-exported from a country that might encounter food supply problems at the same time as other neighbouring economies.
16. When commodities were diverted from developmental food for work projects to emergency operations in Ethiopia in 1984, households dependent on food for work found their food security undermined in a period of continuing drought, and were in turn obliged to seek relief.
17. Clay [*1985*]. Minimum contributions under the FAC are established in wheat equivalent. When these were established in 1980, one tonne of milled white rice was considered equivalent to three tonnes of wheat. That ratio was reduced by agreement to 1:2.4 for 1988/89.
18. This subject was also the focus of a special issue of *Food Policy*, Vol.13 (1) edited by D.J. Shaw and Hans Singer.

REFERENCES

Bachrach, P., 1988, *Opportunities and Risks: An Assessment of the Regular 416 Program*, Washington, Planning Assistance Inc.

Blandford, D. and J. von Plocki, 1977, *Evaluating the Disincentive Effect of PL480 Food Aid: The Indian Case Reconsidered*, New York: Department of Economics, Cornell University, July.

Borton, J., 1989, 'UK Food Aid and the African Emergency 1983–1986', *Food Policy*, Vol. 14, No.3, pp.234–240.

Borton, J. and E.J. Clay, 1986, 'The African Food Crisis of 1982–1986', *Disasters*, Vol.10, No.4, pp.258–272. (Also published in revised version as Ch. 9 of D. Rimmer (ed.), 1988, *Rural Transformation in Tropical Africa*, London: Belhaven Press).

Borton, J., 1986, 'Food aid resource planning: Botswana Experience', paper prepared for the World Food Programme/African Development Bank Seminar (Abidjan, Sept. 1986).

Bryson, J., 1986, 'The Lesotho Food-for-Work Programme of Catholic Relief Services', paper prepared for World Food Programme/African Development Bank Seminar (Abidjan, Sept. 1986).

Clay, E.J., 1983, 'Food Aid: An Analytic and Policy Review (with special reference to developments and the literature since 1977)', Institute of Development Studies, University of Sussex, July.

Clay, E.J., 1985a, 'Review of Food Aid Policy Changes since 1978', *Occasional Paper*, No. 1, Rome: World Food Programme.

Clay, E.J., 1985b, 'The 1974 and 1984 Floods in Bangladesh: From Famine to Food Crisis Management', *Food Policy*, Vol.10 No.3, pp.202–206.

Clay, E.J., 1986, 'Rural Public Works and Food-for-Work: A Survey', *World Development*, Vol.14, No.10/11, pp.1237–1252,

Clay, E.J., forthcoming, 'Poisoned Gift or Resource for Development? Dairy Aid and Developing Countries', in P.P. Anderson (ed.), *Economics of Dairy Development*.

Clay, E.J. and C. Benson, 1988, 'Evaluation of EC Food Aid in Tanzania', Report for EC Commission, London: Relief and Development Institute.

Clay, E.J. and E. Everitt, 1985. 'Food Aid and Emergencies: A Report on the Third IDS Food Aid Seminar', *Discussion Paper*, No.206, Institute of Development Studies, University of Sussex.

Clay, E.J. and J. Pryer, 1982, 'Food Aid: Issues and Policies', *Discussion Paper*, No.183, Institute of Development Studies, University of Sussex.

Clay, E.J. and D.J. Shaw (eds.), 1987, *Poverty, Development and Food*, London:

Macmillan.
Clay, E.J. and H.W. Singer, 1985, 'Food Aid and Development: Issues and Evidence', *Occasional Paper*, No. 3, Rome: World Food Programme.
Clay, E.J. and S. York (eds.), 1987. 'Information and Emergencies: A Report on the 5th IDS Food Aid Seminar 21–24 April 1987', *Discussion Paper*, No.326, Institute of Development Studies, University of Sussex.
Cornelius, J. and M. Letarte, 1987, 'Timeliness of Canadian Bilateral Emergency Food Aid and Deliveries, 1983–1984 to 1985–1986'. Ottawa: Canadian International Development Agency.
European Communities Commission, 1981, 'Food Strategies: a new form of co-operation between Europe and the countries of the Third World', *Europe Information (Development)* DE 40(X374/82) Brussels, Dec.
European Communities Commission, 1982, *Food Aid for Development* (Commission communication to the Council) COM(83)141 final, Brussels, 6 April.
Food and Agriculture Organisation, 1980, *Principles of Surplus Disposal*, Rome.
Hay, R. and E.J. Clay, 1986, 'Food Aid and the Development of Human Resources', *Occasional Paper*, No.7, Rome: World Food Programme.
Hossain, Mosharaff, 1987, 'From Relief to Development: a review of Food-For-Work and Vulnerable Group Development Programmes in Bangladesh', paper for Government of Bangladesh/WFP Seminar on Food Aid for Human and Infrastructure Development in Bangladesh, Dhaka.
Huddleston, B., 1984, 'Closing the Cereals Gap with Trade and Food Aid', *Research Report*, No.43, Washington, DC: International Food Policy Research Institute.
Isenman, P. and H.W. Singer, 1977. 'Food Aid: Disincentive Effects and their Policy Implications', *Economic Development and Cultural Change*, Vol.25, No.2: 205–237.
Kennes, W., 1988, 'Food Aid and Food Security: Approach and Experience of the European Community', *Tijdschrift voor Sociaal Wetenshcapperijk Onderzoek van der Landbouw*, Vol. 3 No.2, pp.138–157.
Maxwell, S.J. (ed.), 1982, 'An Evaluation of the EEC Food Aid Programme', *Commissioned Study*, No.4, University of Sussex: Institute of Development Studies.
Mellor, J.R., 1987, 'Food Aid for Food Security and Economic Development' in Clay and Shaw (eds.), pp.173–91.
National Research Council, 1988, 'Food Aid Projections for the Decade to the 1990s', Washington, DC, 6–7 Oct.
Parotte, J.H., 1983, 'The Food Aid Convention: Its History and Scope', *IDS Bulletin*, Vol.14, No. 2: 10–15.
Relief and Development Institute, 1987, 'A Study of Triangular Transactions and Local Purchases', *Occasional Paper*, No. 11, Rome: World Food Programme.
Relf, C., 1987. 'Food-assisted Rural Employment and Development in Bangladesh: For Institutional Change', paper for Government of Bangladesh/WFP Seminar on Food Aid for Human and Infrastructure Development in Bangladesh, Dhaka.
Singer, H.W., Wood, J. and A. Jennings, 1987, *Food Aid: The Challenge and the Opportunity*, Oxford: Oxford University Press.
Stevens, C., 1979, *Food Aid and the Developing World: Four African Case Studies*, London: Croom Helm.
Stokke, O. (ed.), 1989, *Western Middle Powers and Global Poverty. The Determinants of the Aid Policies of Canada, Denmark, The Netherlands, Norway and Sweden*, Uppsala: Scandinavian Institute of African Studies.
US General Accounting Office, 1986, 'Famine in Africa: Improving Emergency Relief Programmes', *GAO/USAID-86-25*, Washington, DC.
Wallerstein, M.B., 1980, *Food for War – Food for Peace: United States Food Aid in a Global Context*, Cambridge, MA: MIT Press.
World Food Council, 1989, Current World Food Situation, WFC/1989/5, 25 April, Rome.
World Food Programme, 1989a, *Annual Report of the Executive Director: 1988*, WFP/

CFA: 27/P/4, 19 April, Rome.
World Food Programme, 1989b, *Review of Food Aid Policies and Programmes*, WFP/CFA: 27/P/INF/J, 20 April, Rome.
World Food Programme, 1985, 'Evaluation of Food Aid for Price Stabilization and Emergency Food Reserve Projects: A Study of WFP-assisted projects in Tanzania, Botswana, Mauritania, Niger and Mali in 1984', *Occasional Paper* No. 2, Rome.

2

An Additional Resource? A Global Perspective on Food Aid Flows in Relation to Development Assistance

RAM SARAN and PANOS KONANDREAS

INTRODUCTION

The origin of the provision of food aid to developing countries goes back to the 1950s when the United States of America came under pressure to dispose of huge surpluses for which no effective demand existed. For exporters, food aid became a desirable policy choice when the piled up stocks lost most of their market values. In these circumstances, transfer of resources in the form of commodities was seen as largely additional to financial aid. Since the early 1970s, however, world food markets have become tighter and most food commodities are no longer treated as if they were free goods. The establishment of special institutional arrangements for food aid has also raised questions about the concept of its additionality as a resource transfer.

For recipients, food aid has always meant the flow of a resource, providing balance of payments support and budgetary support as well as wage goods needed crucially to keep inflationary pressures under check. But again, whether and to what extent it represents a truly additional resource depends upon the degree to which food aid substitutes for other forms of development assistance. Doubts have also been raised about the cost-effectiveness of food aid received.

Definite answers to the questions and doubts raised above have not

---

Note of acknowledgement

The authors acknowledge useful comments on an earlier draft by Messrs. R.J. Perkins and J. Shaw, and assitance by Mr Di Pancrazio in carryng out the computations involved. The views expressed are those of the authors and do not necessarily represent those of FAO.

been found. A few attempts have been made to explore empirically a number of areas. But conclusions so far reached are mainly conjectural, based on value judgements and often incomplete evidence. This article looks at selected studies and analyses afresh certain aspects of the subject with a view to drawing lessons for future analytical work. Its focus is on methodological issues that are likely to arise in considering the subject of food aid as a resource transfer in terms of presumed additionality, from the standpoint of both donors and recipients.

ADDITIONALITY FROM THE VIEWPOINT OF DONORS

Analysts' Perceptions

From the point of view of donors, food aid can be treated as additional to financial aid if there is no trade-off between the two types of aid. In other words it can be viewed as an extra resource made available by donors if its disbursement is made on its own merits and does not poach resources from financial aid.

Support for Additionality Assumption

As already stated, as long as the source of food aid was burdensome surpluses in major exporting countries, the transfer of food aid was invariably considered to be additional to other forms of development assistance. In the case of the PL480 programme of the United States, the resource cost was shared between a variety of objectives of which managment of domestic surpluses was one; the other important objectives were expansion of export markets and promotion of foreign policy. As Schuh [*1982*] noted, much of the original 'magic' of food aid related to the way it involved resources that had essentially zero value to the donor country. It could therefore be argued that if food exporting donors had not provided food aid, they were not likely to have given additional financial aid [*Clay and Singer, 1985*]. What was true for food aid in cereals in the past became true later for aid in other surplus foods, for example dairy products.

Several other factors also seem to have strengthened the proposition of additionality. It has been pointed out that food aid is a reality – in part because it is easier to get political support for it than for other forms of aid. Support for food aid comes particularly from the farming community – an influential constituency in many countries – as well as from humanitarian and charitable groups. The view that food aid is additional has also received support from the fact that bilateral food aid is funded and administered in some major donor countries both with the involvement of ministries of agriculture and separately from other forms of aid, and that multilateral food aid is provided by the World Food Programme, an agency specially set up for the purpose [*Clay and Singer, 1985*].

Weakening of Support for Additionality Assumption

Notwithstanding the above-mentioned features which support the hypothesis of additionality, the extent to which food aid is to be treated as an additional resource has been called into question in recent years for several reasons. First, during the 1970s world cereal markets were characterised by shortages and interest grew in costing food aid supplies at the price of acquiring the aid commodities on world markets. In more recent years again, the global cereal situation has become fairly easy. But a quantum jump in cereal aid supplies has not taken place although attempts have been made to expand commercial markets through concessional and credit facilities. A view has been expressed that there was neither the political will in donor countries to increase food aid availabilities significantly nor the necessary will in a number of recipient countries to adopt policies and programmes which could use more food aid explicitly for the benefit of the poorest segments of their populations. Another constraint referred to by analysts is the likely adverse effect that these supplies would have on local agricultural production and trade [*Report of the WFP/ Government of the Netherlands Seminar, 1983*].

The gradual diversification of sources of food aid since the late 1960s to include most industrialised countries as well as some others has also weakened the assumption of additionality that had its origins in excess supply problems. Some donors, for example the UK, which did not have any exportable surpluses of food commodities, like cereals, now provide food aid from the same aid budgets from which all types of development assistance are given. The way of funding food aid in other member states of the European Community is about the same. Food aid is an integral part of national aid budgets in all the states except France, so that an increase in food aid – say, for emergencies – would generally result in corresponding cuts elsewhere in the donor's aid programme. Moreover the food aid programmes of individual states known as 'National Actions' (as distinct from 'Community Actions') are funded from national aid budgets rather than from the surpluses resulting from the EC's Common Agricultural Policy [*Singer, Wood and Jennings, 1987*].

The fact that, as signatories to the Food Aid Convention (FAC), major donors provide food aid as a multi-annual commitment is also indicative of the diminishing role of the supply situation in donor countries in determining the volume of food aid [*Singer, Wood and Jennings, 1987*]. It is worth recalling that Australia, which experienced a serious setback to its own wheat production in 1982, partly substituted rice and coarse grains for wheat in its food aid programmes, but did not disrupt food aid shipments during that year. Nevertheless, Australia revised its commitments under the 1980 FAC when it came up for extension in 1986, from 400,000 tons to 300,000 tons. The squeeze on overall budgets in many industrialised countries has also raised concerns about the size, composition and cost-effectiveness of aid budgets.

Surplus Disposal as a Factor Behind Food Aid

The above analysis should not lead to the conclusion that exporting countries no longer experience surplus situations nor implement programmes of surplus disposal and that therefore the validity of the hypothesis of additionality is on the wane. In the case of dairy products, the EC has for long faced problems of surpluses. Accordingly, its food aid programmes provide for large donations in the form of dairy products such as dried skimmed milk, butter and butter oil.

Hopkins [*1983*] has run a simple regression of US wheat stocks and US food aid grain tonnage in the following year to show that the concept of surplus disposal continues to be a working principle of the food aid regime. He states that in the event of unexpectedly large production, Canadians find a food aid channel for rapeseed oil, the Swedes for wheat, the Norwegians for fish, the Germans for egg powder, the Australians and Argentinians for wheat, the Japanese for rice and the United States for a wide variety for products.

FAO has attempted a comprehensive statistical model to assess the degree to which the level of food aid in different cereals of five major donors – Australia, Canada, the EC, Japan and the United States – has been influenced by different factors [*Konandreas, 1987*]. These factors include production levels in donor countries, carry-over stocks held by these countries, world export prices (both current and lagged), production levels in non-donor countries, and food aid commitments under the FAC. The exercise shows that the effect of carryover stocks is substantial for rice (10 per cent, that is an increase by ten tons in food aid for each 100 tons increase in carry-over stocks), moderate for wheat (5.3 per cent) and relatively small for coarse grains (1.1 per cent).

While considering the influence of surplus disposal on food aid, Singer and others remark [*1987*] that, although the importance of surplus disposal as a factor behind US food aid has diminished, surplus disposal has remained important for the US. This is illustrated by the continuing legal requirement in PL480 that there must be a determination of availability by the US Secretary of Agriculture before any shipment of specific commodities can be made under the Act. Singer and others also cite a recent official evaluation of Canadian food aid policy which lays emphasis *inter alia* on the need to take account of Canada's economic interest in increasing the value-added portion of agricultural commodities and reducing the high inventory costs that arise from occasional domestic oversupply of agricultural commodities.

Statistical Relationship between Financial Aid and Food Aid

Singer and Diab [*1988*] have attempted to show that the hypothesis of additionality can be tested by studying empirically the relationship between financial aid and food aid, by adding the value of total development assistance to the other explanatory variables considered in the model by Konandreas referred to above. Overall, their results are inconclusive as regards the additionality hypothesis, reflected by the large number of negative and/or insignificant coefficients obtained. Even when the statistical relationship between individual commodity food aid shipments and total development assistance is positive, this may not be interpreted as representing causality between the two variables. Furthermore, if the aim is simply to check whether a statistical relationship exists between changes in the volume of food aid (real terms as distinct from its nominal value) and changes in financial aid, financial aid should also be included in the model in constant prices. Finally, while introducing development assistance as an explanatory variable, care should be taken to see that it excludes the food aid component; otherwise food aid will figure as both dependent variable and independent variable.

With a view to taking care of some of these deficiencies, an alternative regression analysis has been carried out in this study, between financial aid (excluding the food aid component) and total food aid (including aid in cereals as well as non-cereal foods), both expressed in nominal terms. The results, shown in Table 1, indicate that the relationship was positive and statistically significant for all five donors combined, as well as for individual donors except Japan. This seems plausible because, unlike the volume of food aid in cereals, the nominal value of total food aid has shown a rising trend, albeit modest, thanks to a significant growth in food aid in higher-valued rice, vegetable oils and

TABLE 1

GROWTH RATES OF TOTAL VALUE OF FOOD AID AND FINANCIAL AID AND RELATIONSHIP BETWEEN THEM (1969–86)[1]

Donor	Annual Growth Rates (%)		Regression Coefficient Between Food Aid and Financial Aid[2]	R^2
	Food Aid	Financial Aid		
Australia	12.0	10.7	0.114 (4.21)	0.53
Canada	7.6	11.6	0.156 (6.17)	0.71
EEC	14.2	11.7	0.065 (9.74)	0.86
Japan	2.2	17.7	0.011 (0.77)	0.04
USA	4.4	9.0	0.115 (5.05)	0.61
All DAC[3] countries	6.7	11.7	0.074 (10.96)	0.88

1. Both food aid and financial aid are expressed in nominal value terms.
2. Food aid is taken as the dependent variable. Numbers in parenthesis are t-ratios.
3. Member countries of the OECD Development Assistance Committee.

dairy products, and non-food costs, which correlates positively with the rising trend in financial aid.

In view of the positive value of the estimated correlation coefficients, it can be stated that, generally speaking, in years when the volume of food aid increased (either to respond to increased needs of recipients or to ease the burden of surplus supplies in the donor countries) there has not been a corresponding decrease in the availability of financial aid. Thus, food aid appears to be an additional resource in the short term although in some years movements in food aid and financial aid may not be in the same direction. However, this does not necessarily guarantee additionality over the longer term. As shown in Table 1, food aid and financial aid have both grown over time though not at the same rate, providing some *prima facie* evidence to support the hypothesis that food aid is additional to financial aid. But a closer scrutiny of the data shows that in real terms food aid from Japan fell and from the USA perhaps remained more or less stagnant, while financial aid from both countries increased at significant rates. Should these trends be interpreted as indicating an increasing trade-off between food aid and financial aid? Statistical evidence does not provide a definite answer to this question. It requires a detailed examination of the factors that influence the volume of food aid given by donors. The following section attempts to analyse some of these factors.

Opportunity Cost of Providing Food Aid

We have seen that there has been a tendency to cost food aid at different values depending on the supply/demand situation. Thus when the world supply situation was characterised by burdensome surpluses, food aid was considered to be almost a free good and was therefore treated by exporter-donors as an additional resource which may not have reduced resources for the financial aid programmes. But later, when the supply situation became tight, food aid was costed at world market prices, thus calling into question its additionality as a resource. This way of costing food aid is subjective and arbitrary. Thus, an objective approach is needed which would reflect the real cost or opportunity cost of food aid.

The opportunity cost is expected to provide a more reliable indicator of the sacrifice involved on the part of donors. In the absence of precise information on the motives behind the implementation of food aid programmes, the opportunity cost can also provide at least some insight into the decision-making which affects the level of food aid, given that there is a choice between food aid and financial aid. When the real costs of food aid are high, it is equivalent to financial aid. On the other hand, when real costs are low, it becomes the obvious alternative if a donor has to choose between food aid and financial aid; at times these real costs may fall so low that food aid may provide the most viable method of coping with supply–demand imbalances in food-exporting countries. Since real costs are assumed to be an important factor behind food aid policy decisions, an attempt is made in this article to estimate these costs and explore whether food aid flows in recent years have reflected them.

In the case of food aid, the opportunity cost is the amount that could be earned by selling the commodity in the commercial market rather than providing it free or on concessional basis. Given the fact that recipients of food aid are by and large low-income countries whose capacity to import commercially is severely constrained by lack of foreign exchange while the level of supply is determined exogenously by domestic production policies in exporting countries, a reduction in the volume of food aid and the resulting increase of supplies available for sale commercially are likely to have the effect of depressing market prices.

The Annex provides a conceptual and analytical framework which makes it possible to assess the impact of a hypothetical withdrawal of food aid on world wheat prices and consequently on donors' (exporters') total revenue. The approach followed is highlighted in Figure 1, which depicts aggregate market supply (S) and demand (D) facing the five major wheat exporters when food aid is provided, resulting in an equilibrium price P. Under the hypothetical situation of complete withdrawal of food aid, the supply curve shifts to the right to S' by the amount of food aid withdrawn, and the demand curve also shifts to the right to D', but by an amount generally less than the food aid withdrawn. The extent of the shift of the demand curve depends on the degree of substitution of food aid for commercial imports. Because demand shifts less than the supply curve, the resulting price (P') will be less than that prevailing when food aid was given. It is clear that, since both the quantities sold commercially and the prices for which they are sold are different under these two situations, revenues will generally also differ. In a food aid regime, total revenue is represented by the area of the rectangle OAEP, whereas when food aid is not given total revenue would be represented by OA'E'P'. Thus the opportunity cost of providing food aid will be represented by the difference in areas of the two rectangles OA'E'P' and OAEP.

The analysis in the Annex shows clearly that the opportunity cost to donors from the provision of food aid in wheat depends crucially on the extent to which food aid substitutes for commercial imports. A review of some studies which have attempted to estimate this coefficient is presented in a subsequent section. It follows that the degree of substitution depends on the period for which estimates were made, reflecting the shift in the set of countries receiving the bulk of food aid. Overall, based on this review as well as other considerations, a coefficient of substitution equal to 0.6 is assumed here (see below, p. 54). Results obtained for other values of this coefficient are given in the Annex.

The analysis is restricted to wheat, which accounts for about 70 per cent of food aid in cereals, and it covers the 17-year period from 1971/72 to 1987/88. The results of the analysis show that if donors had not provided food aid in wheat, world prices of wheat between 1971 and 1988 would have averaged US$136/ton, that is, about US$3 lower than the actual average price of US$139/ton (when food aid was given). The average revenue to donor exporters if they had hypothetically

FIGURE 1
EFFECTS OF ELIMINATION OF FOOD AID ON WORLD PRICE

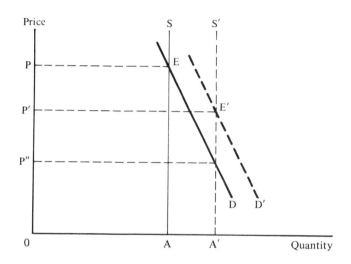

withdrawn all food aid in wheat would have amounted to US$21,750 million compared to US$21,586 million when food aid was given. Thus, the loss of revenue to donors was only US$174 million. Expressed on a per unit basis, the real cost of food aid amounted to only US$22/ton. To this should be added the cost due to an increase in consumer food prices in donor exporting countries resulting from an increase in world prices. Taking this factor into account, the real cost of food aid would have increased to US$54/ton.

Several other factors, which would also affect the real cost of food aid, were not considered in the model. One was that, at the new equilibrium price of US$136/ton (when food aid was not given) there may have been a marginal increase in demand because of the price elasticity effect, as a result of which the revenue earned would have been somewhat higher. Second, as the coefficient of substitution of food aid for commercial imports was assumed to be 0.6, 40 per cent of the average level of food aid in wheat (of 7.2 million tons), amounting to about three million tons, would have to have been carried over to the following year with a further depressing effect on prices in that year. This would be a continuing process and the cumulative effect on prices and revenues could be substantial. In addition, in this case, donors would also incur higher storage costs for carrying a larger amount of stocks over time. Both these factors imply a decrease in the real cost of food aid.

Despite its limitations, the above approach has merit. It has given at least a rough idea of the true cost of food aid. There is, however, the need to refine the model and to improve its value as a tool for more appropriate valuation of food aid. It is also appropriate to extend the commodity coverage of the analysis. It should also be pointed out that the calculations presented do not take into account some other donor costs associated with the provision of food aid (that is, ocean transport and in some cases internal transport and handling in the recipient countries), nor the terms of concessional sales. These are briefly elaborated below.

Other Considerations Relating to the Real Cost of Food Aid

In estimating the opportunity cost of food aid to donors, consideration needs to be given to at least two additional factors. First some donors (mainly the United States and Japan) have been providing a large volume of food aid on concessional or easy credit terms. Through payments over an extended period, donors recover a part of the cost of food aid given in this form. The discounted value of the payments that recipient countries must eventually make has been estimated at US$416 million a year for food aid in cereals for the period 1976–78 [*Huddleston, 1984*]. On average this comes to US$50 per ton which has to be set against the net cost that donors incur in the provision of food aid (see also p.50ff for further information).

The second factor which has the opposite effect, that is, increasing the real cost of food aid to donors, is the need to finance (i) procurement, transport and delivery to the point of loading in the exporting country, (ii) transport from the port of loading to the delivery point in a recipient country, and (iii) for some recipients, meeting internal transport costs to the final destination. The first category of operations is the major responsibility of donors and costs on this account are included in the f.o.b. price. Donors also bear in most cases the cost of the second operation involving overseas shipments, though in a large number of cases recipients are required to pay this. In 1982/83 freight charges paid by donors amounted to US$344 million. Donors other than the US paid about US$100 million for shipping 1.8 million tons of grain out of a total of 2.8 million tons provided by them according to the International Wheat Council [*1984*]. The US paid US$246 million but

AN ADDITIONAL RESOURCE?

the requirement that half of all government-generated cargoes should be shipped by relatively uncompetitive US flag carriers inflated the cost of transport. It has been estimated that an extra US$100 million has to be spent under this requirement [*Minear, 1983*]. Thus the real cost of overseas transport financed by donors may be estimated at about US$250 million, or say an average cost of US$25 per ton.

The above analysis serves to illustrate that the opportunity cost of a commodity can be quite different from the assumed cost based on subjective impressions. It has been shown that during 1971–88 aid in wheat had the effect of raising its market price by an average of about US$3 per ton (2.2 per cent) which sellers in exporting countries realised for the entire quantity marketed, and which contributed to a sharp reduction in the resource cost of wheat food aid – from US$139/ton (apparent gross cost) to US$54 per ton (opportunity cost). Moreover, the provision of food aid in cereals on concessional terms resulted in the recovery of over US$400 million (reducing the cost of cereals by US$50/ton between 1976 and 1978). Gains due to these two factors would, however, be offset by the payment made by donors for the shipment of food aid commodities (in 1982/83 it amounted to about US$25 per ton).

From the above it appears that the opportunity cost of food aid in wheat during 1971–88 was hardly one-fourth of its average international price. This represents the situation at the global level. The value of the real cost, however, is expected to be different for different donors depending upon the volume of their commercial exports which benefit from higher prices realised when food aid is given, the incidence of the additional burden that consumers in these countries have to bear as a result of higher prices, the volume of food aid given on credit, and transport and other non-food costs borne by them. It is true that those countries whose commercial exports are large in relation to the volume of food aid, bear a relatively small real cost of providing food aid. Similarly, only those countries which provide food aid on credit recover a part of its cost. Non-food costs of food aid borne by some donors have increased in recent years but these have remained relatively small for others. Thus the real cost of food aid for each donor should be determined after taking into account all quantifiable gains and losses that may result from the provision of such aid.

Conclusions from the Viewpoint of Donors

The objective of the above analysis was not to provide precise data on the opportunity cost of food aid but to suggest a methodology which would give a broad idea of the sacrifice involved on the part of donors in implementing food aid programmes and improve our understanding of the additionality or otherwise of food aid as a resource transfer. The main finding of the analysis that has a bearing on additionality is that since the beginning of the 1970s, when the cereal supply situation has been relatively tight, the opportunity cost of food aid has been much below the cost of acquiring the aided commodities on the world market. Moreover, estimates of the resource cost could be depressed further considering that food aid also often serves the objectives of market expansion and foreign policy promotion. In the light of the information on the strikingly low levels of real costs, a question may however be asked: why has food aid not expanded as rapidly as it did in the 1950s and 1960s? Is it because food aid flows are constrained by financial factors?

To put the record straight, let us first look at the recent trend in food aid shipments. Between 1984/85 and 1987/88, when exporting countries experienced surpluses, food aid did rise at least modestly. Cereal aid averaged 12 million tons in these four years compared to only nine million tons in the preceeding four years. However, again in 1988/89, when drought has affected crops in important donor/exporting countries, shipments are expected to drop to below 10 million tons.

One could argue that growth in food aid has possibly been constrained by limited absorption capacity in recipient countries. But this argument is indefensible because these countries have continued to spend a large part of their scarce foreign exchange resources on the commercial import of food commodities. One could offer two further explanations. One is that perceptions of real costs remain vague until countries face huge surpluses over prolonged periods. In the case of cereals, undoubtedly surpluses have appeared in recent years as a result of production policies pursued by major exporters, but so far the situation has been dealt with largely through stock accumulation and sales through various subsidised programmes. On the other hand, in the case of dairy products, which have been in surplus for long periods, real costs appear to have been taken into account in disposing of surplus quantities by expanding food aid programmes.

The second possible explanation for a restrained growth in food aid could be the lack of a coordinated approach among major donor/exporters to find solutions to the problem of surpluses. Unlike in the past when only one major exporter had huge surpluses, now several producing countries face supply–demand imbalances. Consequently, any one country attempting to deal with the problem of surpluses through unilateral action is likely to bear a disproportionate share of the cost of adjustment, although such an action would have benefits for all the exporters. At the same time, there is a growing realisation that an expanded food aid programme is expected to provide a relatively cheaper method of coping with production surpluses than other methods such as crop area reduction schemes or stock accumulation. The fact that donors increased food aid shipments in recent years when they faced strong growth in production in the face of slowdown in the expansion of demand, reflects the changing perception about the role of food aid in disposing of surpluses. But whether the amount of food aid given by a country is determined more by its food supply situation or by its financial situation, will remain blurred until food aid programmes grow substantially as a result of either a multilateral agreement or unilateral action on the part of individual donors. But for those countries, like the United Kingdom, which have no exportable surpluses, any increase in food aid will no doubt be at the cost of development assistance in other forms.

In the light of the above, it should be concluded that it is not easy to make a judgement on whether food aid as a resource is additional to financial aid or not. A study of real costs would certainly help towards a better understanding of the situation. Together with that, it would be necessary to analyse all other factors which have a bearing on government food aid policies. A rigorous analysis would be needed of all relevant factors, including the budgetary procedures followed in donor countries, particularly if food aid remains stagnant while real costs are low.

ADDITIONALITY FROM THE VIEWPOINT OF RECIPIENTS

Recipient countries accept food aid because it provides them with much-needed food to feed their poor and hungry people. To them food aid, as a resource, appears additional to financial aid because they are not sure if, in its absence, they would receive financial assistance of an equal or even somewhat smaller value. The situation in most cases is not one of choosing between food aid and the same amount of financial aid, but rather between food aid and nothing [*Singer, Wood and Jennings, 1987*]. The question that concerns these countries in regard to the presumed proposition of additionality is whether the transfer of food as a resource is as efficient and valuable to them as financial aid is.

The question raised above needs to be examined in relation to the support that food aid offers to recipient countries in accelerating the process of development. Such support comes by giving these countries an opportunity to use for development the foreign exchange that may be set free and the additional domestic resources that may be generated by the availability of food aid. Therefore the answer to the question hinges mainly on the examination of two issues: (i) the contribution that food aid makes to the reduction of the foreign exchange burden of recipient countries, and (ii) the amount of budgetary and other resources that food aid provides for investment.

The contribution that food aid makes to the relief of foreign exchange burdens can be assessed in a wider, as well as in a narrow, sense. In the wider context, it amounts to determining the true value of the entire food aid package, that is, the value expressed in a foreign currency such as dollars after allowing for any costs that recipient countries may bear in receiving the aided commodities. But food aid donations consist broadly of two parts, one which substitutes for commercial imports and the other which results in additional consumption and does not displace commercial imports. The first part is nearly equivalent to a gift of foreign exchange that can be used for the import of other essential items. But the same cannot be said about the second part. In the narrow context, the assessment of the contribution of food aid should be confined to only that part which substitutes for commercial imports.

The True Value of Food Aid to Recipients

Determination of the true value of food aid to a recipient country involves an estimation of the foreign exchange it would have spent to import an equivalent quantity of food purchased on the world market. Food aid statistics compiled by the OECD Development Assistance Committee are broadly based on the cost price of food to the donor and/or commercial prices [*Singer, 1987*]. But these figures inflate the value of food aid because:

(i) as mentioned earlier, a part of the cost of food aid received on concessional terms has to be repaid;
(ii) when food aid is tied to purchases from donor countries, a part of it may be valued at a price which is higher than world market prices;
(iii) the face value of food aid does not take into account the possibility that world market prices may have been lower if food aid had not been provided;
(iv) recipients would not have imported some of the exotic foods donated or would have imported cheaper foods to save foreign exchange.

AN ADDITIONAL RESOURCE?

Analysts have attempted to discount the recorded value of food aid in the light of some of the above-mentioned factors. Schultz, in his pioneering article nearly 30 years ago [*1960*], addressed this very question and after discounting for several factors came to the conclusion that the value of PL480 products to the countries receiving them was perhaps as low as 37 cents for each dollar of food aid as accounted by the Commodity Credit Corporation (CCC). More recently, Huddleston [*1984*] considered the value of food aid to represent the difference between the c.i.f. value of food aid imports and the true cost of these imports which recipient countries must pay from their own foreign exchange resources. As mentioned earlier, it became necessary to determine the true cost of food aid imports because a substantial part of aid is provided on easy credit terms and has both a grant element and a commercial element. Huddleston has worked out the value of the grant equivalent of these concessional imports and added it to the value of pure grant aid imports to arrive at an estimate of the contribution that food aid makes to reducing the foreign exchange burden of recipient countries. Between 1976 and 1978 the total annual contribution was estimated at US$841 million, that is, two-thirds of the c.i.f. value of US$1 257 million. The balance of about US$400 million represented the foreign exchange cost borne by recipient countries themselves.

The estimates provided by Huddleston have the advantage of being based uniformly on the c.i.f. value of cereal imports rather than on the cost price of food to donors, or some other market price. But these estimates still tend to exaggerate the true value of food aid to recipients because a significant part of food aid is provided on f.o.b. basis rather than c.i.f. basis. Huddleston's estimates will therefore have to be discounted to allow for the overseas transport costs borne by recipients. Data on shipping coasts available from the International Wheat Council can be used to arrive at estimates of transport costs paid by recipients. It is also necessary to make an allowance for the extra amount (about US$100 million) which has to be spent on shipping half of the US goverment-generated cargoes by relatively uncompetitive US flag vessels [*Minear, 1983*].

It was seen earlier that world market prices of wheat could possibly have been lower by about US$3 per ton if food aid had not been provided. Recipients would have paid a lower price for the entire quantity imported commercially. The gain to recipients on this account would have been substantial because commercial imports account for the bulk of developing countries' cereal imports. Low-income, food-deficit countries, which received on average about seven million tons of wheat as food aid during the 1980s, also imported commercially an average of about 28 million tons a year. Thus, if allowance is made for the higher price paid by recipients for their commercial imports when food aid programmes are implemented, the true benefit of food aid would be about US$85 million less than appears to be the case.

For the purpose of obtaining true values, food aid in exotic commodities such as dairy products, and wheat and rice for some countries, should normally not be valued at the high cost price to the donor or at world market prices. Dairy products are luxury items, consumed largely by better-off people. There is little likelihood of low-income countries importing them commercially if food aid were not provided. Food aid in wheat and rice received by those countries whose staple food is coarse grains should normally be valued at prices at par with coarse grain prices. But tastes for wheat and rice have already been developed in many recipient countries. Determining the imputed values of these commodities is therefore a matter of value judgement for an analyst undertaking a study of food aid benefits.

The Impact of Food Aid on Concessional Imports

The question of substitution of food aid for commercial imports has attracted the attention of many analysts ever since the launching of the food aid programmes in the 1950s. It has always been recognised that such substitution amounts to a gift of foreign exchange to recipient countries and provides balance of payments support just as financial aid does [*Clay and Singer, 1985*]. However, concerned with the harmful effects that concessional sales can have for the commercial trade of exporting countries, the international community has agreed on the FAO Principles of Surplus Disposal. These represent a code of conduct to be observed by governments exporting food and agricultural commodities on concessional terms and stipulate that food aid can only be additional to, and not a substitute for, commercial imports [*FOA, 1980*]. The Principles seek to ensure that concessional sales result in additional consumption in recipient countries and do not displace normal commercial trade. The mechanism for assuring such additionality is the UMR (Usual Marketing Requirement) provision in food aid agreements, which requires a recipient country to maintain at least a specified minimum level of commercial imports in addition to concessional supplies. Normally a statistical figure representing the actual commercial imports over the preceding five years provides the basis for arriving at the UMR. But there is also a provision for modifying this figure, taking into consideration a number of other factors such as the economic and balance of payments position of a recipient country and its development needs. For most low-income countries, scarcity of foreign exchange has been an important factor in determining the level of UMR. The application of the UMR instrument is therefore very likely to have led to at least some substitution of commercial imports by food aid.

This conjecture is confirmed by many analysts. In estimating the net import demand of 16 food aid recipients, for two commodities each, Abbott [*1979*] found evidence to support the presumption that the constraint of additionality of consumption was circumvented in 26 out of 32 cases. An impact evaluation of the PL480 Title I programme for Sri Lanka [*USAID, 1982*] also acknowledged that food aid under this programme represented a balance of payments resource over the 25 years of its existence, implying that food aid substituted for commercial imports. Examining the cases of those countries which faced a cutback in food aid between 1964 and 1978, Braun and Huddleston [*1986*] came to the conclusion that only one-half to two-thirds of them substituted commercial imports for the lost aid in the short run. Thus, even though in theory food aid should not substitute for commercial imports, in practice receipts of food aid have been offset by reduced commercial imports. The main question to be considered now is: by how much does this substitution occur?

Review of Evidence on the Substitution Effect

The conclusion reached by Clay and Singer [*1985*] on the basis of studies carried out by different analysts is that food aid *de facto* substitutes to a significant degree for commercial imports in a number of important importing countries such as Egypt, Sri Lanka and the Republic of Korea; whereas in India less than a quarter of cereals food aid substituted for commercial imports. A consensus seems to be emerging among analysts that food aid for sale on the domestic market is a substitute for commercial imports. EC cereal food aid provided for sale, which accounts for about a quarter of the Community programme by value, is in effect considered to provide balance of payments support [*IDS and Africa Bureau, Cologne, 1982*]. Maxwell and Singer [*1979*] estimated that most food aid, about two-thirds, is sold by recipient countries on their domestic markets to extend supplies and generate funds. It is contended that not all of this need substitute for commercial imports, but evidence from four case studies on Botswana, Lesotho, Tunisia and Upper Volta [*Stevens, 1979*] suggests that a significant part of it may do, and also that some project aid has the same effect.

With the aim of quantifying the relationship between imports and food aid Braun and Huddleston [*1988*] fitted a simple regression with deviation from trend of food aid as explanatory variable and deviation from trend of commercial imports as dependent variable using data for 1962–78. The equation showed that for 13 of the 32 countries a reduction of food aid by one ton from trend levels increased commercial imports by 0.8 tons. However, the equation did not consider a number of variables which influence commercial imports. Given the simplistic approach, the inference drawn from these results is that substitution occurs to some extent, but that food aid cannot be viewed either as simple balance of payments support, which would be the case if it substituted completely for commercial imports, or as fully additional.

In an analysis of the impact of food aid on commercial imports, Pinstrup-Andersen and Tweeten [*1970*] came to the conclusion that, if no wheat had been available under food aid programmes, the survey countries would have imported 41 per cent of the wheat they received as aid. The basic data on commercial imports replaced by food aid was obtained through a mailed questionnaire circulated among 441 persons representing 44 countries. A partly or fully completed questionnaire was received from 88 of the 441. Although this information had important policy implications at the time of the survey, it is considered outdated for various reasons: in the 1960s several producing countries had huge surpluses of wheat; most of the developing countries covered by the survey receive at present marginal quantities of food aid. However, even though the results of the survey are no longer strictly valid, they do provide a broad indication of the impact that food aid can have on commercial imports.

An Alternative Approach to Assessing the Substitution Effect

In view of the problems encountered with *ad hoc* attempts to assess the degree of substitution of food aid for commercial imports, this paper puts forward an alternative methodological approach, by extending a model developed by FAO to assess food aid requirements for 1985 [*FAO, 1983*]. FAO computed elasticities of commercial cereal import demand with respect to the ratio of export earnings to cereal import prices on a country by country basis; used these elasticity coefficients to estimate commercial imports for the year 1985; and arrived at estimates of food aid requirements, representing the difference between estimated total requirements and estimated commercial imports for 1985. The elasticities computed by FAO (using data for the period 1970 to 1981) ranged between 0.2 and 1.2. They can be used to determine the level of commercial imports that food aid recipients would have to arrange if food aid were no longer provided. The specification of the

AN ADDITIONAL RESOURCE?

additional commercial import demand would relate the quantity of cereal imports to the hypothetical level of foreign exchange involved in the receipt of non-project food aid by recipient countries and the per unit price of cereal imports. Based on the FAO model, the theoretical relationship to estimate commercial cereal imports in the absence of food aid would have the following form:

$$I = a(F/P)^b$$

where I = projected commercial cereal imports;
F = the foreign exchange value of non-project food aid;
P = unit price of cereal imports;
a = the constant term of the estimated relationship; and
b = the estimated elasticity of cereal import demand with respect to the ratio of export earnings to cereal import prices.

The export earnings variable in the FAO model (made up of merchandise, services and private transfers) has been substituted by F (that is, the foreign exchange value of non-project food aid).[1]

The above is the first stage in assessing the degree of substitution of food aid for commercial imports through a comprehensive model. For this purpose, in the context of the approach suggested here, new factors will also have to be considered. They include a reassessment of domestic cereal production and effective demand, taking account of a reduction in domestic resources resulting from the termination of food aid, and new priorities in the allocation of foreign exchange for the import of cereals and other essential items. Data constraints could also pose a problem for a country by country application of this model. However, a beginning could be made by using the estimated parameters in the FAO study with suitable modifications.

The above approach would provide estimates of the volume of commercial imports which recipient countries may need to arrange in the absence of non-project food aid, that is, the aid which is provided for sale. In addition, food aid is provided in the form of project aid and emergency aid. The project aid is designed to result in additional consumption and generally does not provide balance of payments support. Therefore, when food aid is terminated, projects financed with food aid are likely to be axed for reasons of balance of payments difficulties. Emergency food aid imports are required in situations caused by natural and man-made events, and will therefore continue to be needed.

Thus, in estimating the level of imports in the absence of food aid programmes the ratio of substitution of non-project food aid presents the greatest problem. For working purposes it can be assumed that, in the absence of food aid, commercial imports would be able to meet 60 to 65 per cent of the *effective* demand gap, and the remainder would remain unmet.[2] This assumption and the reasons given for the continuation of emergency aid, indicate that commercial imports are likely to substitute for about 60 per cent of total food aid now received.

Generation of Domestic Resources

It is recognised that resources generated by the sale of food aid commodities offer the recipient an additional resource for development. Under the food aid agreements between donors and recipients, sales for local currency are generally permitted only of those commodities which are received on concessional terms, and many governments sell these at subsidised prices. A view is now being expressed that such generation of counterpart funds is a misrepresentation of the role food aid can play in promoting development, because if the products are sold with a subsidy the value of the counterpart funds generated may be less than the value of the food aid resource. In that case the size of the counterpart funds may inadequately represent the contribution that food aid makes to development [*IDS and Africa Bureau Cologne, 1982*]. It has been suggested that the role of food aid as a resource transfer should be seen in a macro-economic sense, giving due consideration to its effects on savings, investment and balance of payments. It is difficult to assess this wider impact, but it should be recognised that food aid can be seen as providing resources not only when commodities are sold but also when they are used for other purposes, for example, financing projects. In this sense, almost all food aid should be viewed as representing an additional flow of resources for the domestic economy.

Conclusions from the Standpoint of Recipients

For recipients, food aid has always meant the flow of an additional resource, but the foregoing analysis raises questions about the extent to which it is truly additional. Many recipients have to make payments for food aid received on credit terms as well as for meeting the overseas transport cost for a significant number of shipments. Also, the face value of food aid does not take into account the possibility that world market prices may have been lower if food aid had not been provided. To arrive at the true value to recipients, an allowance has therefore to be made for various payments they make in receiving food aid as well as for the higher prices they pay for commercial imports because of the implementation of food aid programmes. An attempt to do this shows that the true value of food aid is likely to be much below, perhaps half of, its face value. However, there are substantial differences among food aid recipients. Major net beneficiaries are those countries whose imports consist largely of food aid, which receive food aid mostly on a grant basis, and which are given significant amounts of non-food aid along with food aid. On the other hand, the true value of food aid is lower for those countries whose imports are weighted in favour of commercially procured food (rather than food aid) or concessional imports (rather than pure grant imports) or which have to bear a significant part of overseas transport costs.

In analysing food aid as a resource transfer to recipients, it has also to be recognised that a significant part of food aid results in additional consumption in these countries and does not release foreign exchange which could be used for the import of other essential items. Only that part of food aid which substitutes for commercial imports provides balance of payments support. Attempts have been made in the past to quantify this part of food aid, but a fresh look is needed at methodological questions. A methodology is considered here which uses information on elasticities of commercial cereal import demand with respect to the ratio of export earnings to cereal import prices on a country by country basis. Such elasticities have been computed by FAO, with the objective of assessing food aid requirements.

The role of food aid in generating resources through the sale of aided commodities is also of crucial importance for many recipient countries. Collection of information on these resources requires the help of donors and recipients. In a wider sense, all food aid can be interpreted as representing an additional flow of resources available for development; the collection of information on the volume of such aid presents fewer problems.

ANNEX

ANALYTICAL FRAMEWORK FOR ESTIMATING REAL COST OF FOOD AID

(a) Determinants of market-clearing price with and without food aid

Crucial in the analysis and assessment of the true cost of food aid to the donor country is its impact on the market price. The starting point in determining the market price of any commodity is to define the relevant market where that price is formed. In the case of world wheat prices, the dominant role of the five major exporters (USA, EC, Canada, Australia and Argentina, accounting for as much as 95 per cent of world exports of wheat) suggests clearly that the relationship between supply and demand for these five exporters should be largely the determining factor. Figure 1 (see main text) represents graphically this relationship during a given year, where the vertical line at point A represents aggregate market supply and D represents total *market* demand (domestic and export) facing the five major exporters. P represents the equilibrium price. The stress on *market* supply and demand is to highlight the assumption that quantities provided as food aid are removed from the total supplies before the market-clearing price is determined, that is, they are not part of the commercial market, implying that quantity OA is equal to carry-over stocks plus current production plus imports minus food aid.

Now assume that food aid is no longer provided. This has both supply-side and demand-side effects. First, an amount equal to the level of food aid (F) is added to the supply, shifting the market supply line to the right, from A to A', where AA' = F. Second, on the demand side, the elimination of food aid would be reflected by an increase in total commercial demand by an amount αF, where the substitution coefficient should observe the relationship $0 < \alpha < 1$. The new equilibrium price P' will be as low as P" if there is no shift in the demand curve ($\alpha = 0$), that is, if the food aid provided was 100 per cent additional to commercial imports, and as high as P if the demand curve shifts by an amount equal to the reduction of food aid ($\alpha = 1$), that is, if there was no additionality of food aid to commercial imports. In the most likely cases P' will be somewhere between P and P".

(b) Estimation of market-clearing price

The above exposition represents a standard model for determining the price of a commodity in a comparative market environment. Equilibrium price is determined by the intersection of supply and demand, and, starting from an initial equilibrium price P it is possible to express analytically the new equilibrium price P' on the basis of the initial quantities and the elasticity of the aggregate demand. This approach was followed, *inter alia*, by Pinstrup-Andersen and Tweeten [*1970*] in determining the possible effect of food aid on the world price of wheat. While analytically attractive, this approach requires knowledge of relevant elasticities and assumes linearity of the demand curve. Estimates of price elasticities of demand (domestic and foreign) for individual commodities have been made by various analysts; however, the range of elasticity values obtained is quite wide and their statistical significance is often low.

An alternative to the conventional approach is to attempt to estimate directly a price relationship on the basis of a composite variable that could best be expected to reflect *ex post* the degree of relative scarcity or relative abundance of a commodity in the market. In the case of wheat, such a composite variable, it is argued, could be represented by the ratio of total *market* supplies and total *market* demand during a given marketing year, that is,

$$R_t = \frac{S_t + O_t + I_t - F_t}{U_t - F_t}$$

S_t = opening stocks of the five major exporters in year t,
Q_t = current production of the five major exporters in year t,
I_t = imports of the five major exporters in year t,
U_t = total utilisation (domestic + exports) by the five major exporters in year t, and
F_t = food aid provided during year t.

In a situation of low total supplies and/or high total utilisation, resulting market prices would be high and *vice versa*. Thus, the expected coefficient between R_t and P_t is expected to be negative. The specific functional form that links the two is an empirical question. After some experimentation with different alternatives, a logarithmic specification was considered as the most appropriate. The estimated relationship for wheat using data from 1971/1972 to 1987/88 is as follows:

$$\ln P_t = 5.35 - 3.09 \ln R_t + 0.25 \ln t \qquad R^2 = 0.69$$
(t-ratio)   (–4.19)   (4.46)

The time trend was incorporated into the relationship to capture any trend in the price variable which could not be attributed to the basic explanationary variable R_t. Both coefficients of the explanatory variables are highly significant and the explanatory power of the estimated relationship is satisfactory. The estimated coefficient of the R_t variable represents directly the relevant elasticity, that is,

$$\frac{\Delta P_t / P_t}{\Delta R_t / R_t} = -3.09$$

The latter relationship represents an operational formula by which it is possible to estimate changes of price levels due to exogenous shifts in either total supply and/or utilisation during a year. Thus, the new equilibrium price (P'_t) that would result from an assumed elimination of food aid will be given by the relationship:

$$\frac{P'_t - P_t}{P_t} = -3.09 \frac{R'_t - R_t}{R_t}, \text{ or}$$

$$P'_t = P_t \left(1 - 3.09 \frac{R'_t - R_t}{R_t}\right)$$

where $R_t = \dfrac{S_t + O_t + I_t - F_t}{U_t - F_t}$, $R'_t = \dfrac{S_t + O_t + I_t}{U_t - (1-\alpha)F_t}$ and

α = coefficient of substitution of food aid for commercial imports.

(c) Estimation of real cost of food aid to donor countries

As discussed above, the market equilibrium in the two contrasted situations, that is, with and without food aid, is different both in terms of market-clearing price and the level of commercial utilisation. Although market sales are generally lower with food aid than without, market prices are higher, so that the cost (or gain) of providing food aid can be assessed by a comparison of the relevant revenues in the two situations, that is, cost of food aid will be equal to the revenue without food aid minus the revenue with food aid.

The difference in revenues to be considered, corresponds to the total commercial utilisation (domestic and exports) of supplies of wheat during a particular year. Analytically, the real cost, expressed per unit of food aid provided, amounts to:

$$RC_t = \frac{\left[U_t - (1-\alpha) F_t\right] P'_t - (U_t - F_t)P_t}{F_t}$$

where the nominator represents the difference in revenues between the two situations.

This cost is expressed as a function of the coefficient of substitution α between food aid and commercial exports. It should be expected that, at the one extreme, when $\alpha = 0$, that is, when elimination of food aid does not result in any increase in commercial imports (or conversely when food aid is totally additional to a recipient country's commercial imports) the cost to the donor country of providing food aid would be the least (or the benefit the highest), as in this case donors benefit from the higher price resulting from the provision of food aid while at the same time maintaining commercial sales. On the other hand, at the other extreme, when $\alpha = 1$, that is, when there is complete substitution between food aid and commercial imports, the provision of food aid reduces ton-per-ton the recipient country's commercial imports, so that there is no increase in price, while part of the supplies that previously were sold commercially are now given free in the form of food aid. Thus, in this case the cost of food aid to the donor country is expected to be the highest (or the benefit the lowest).

These expected results are confirmed by application of the above derived relationship for the period from 1971/72 to 1987/88. For a given value of the substitution coefficient α, year-by-year costs were computed which were then averaged for the whole 17-year period. The outcome is shown in Table 2. It turns out that food aid in wheat represents a cost to the donor countries if the coefficient α is greater than 0.56, that is, if the provision of one ton of food aid results in a reduction in commercial demand by more than 0.56 tons. If the reduction in commercial demand is below this level, then provision of food aid actually represents a net benefit to donor countries.

The consumer effect of higher prices resulting from the provision of food aid need also be considered. To the extent that such higher world prices are reflected in correspondingly higher prices paid by

TABLE 2
ESTIMATES OF REAL COST TO DONORS OF PROVIDING FOOD AID, UNDER DIFFERENT VALUES OF THE SUBSTITUTION COEFFICIENT α (1971–88)

Coefficient of substitution of food aid for commercial imports (α)	Average annual food aid before withdrawal	Average annual total commercial utilisation		Average world price		Average annual total revenue		Real cost of providing food aid		Increase in consumer expenditure when food aid is provided	
		with food aid	without food aid	with food aid	without food aid	with food aid	without food aid	Average annual	Per unit of food aid	Average annual	Per unit of food aid
	(... million tons ...)	(... million tons ...)		(... US$/ton ...)		(... US$ million ...)		(US$ million)	(US$/ton)	(US$ million)	(US$/ton)
0.3	7.2	153.5	155.7	139	131	21586	20582	−1004	−150	685	103
0.4	7.2	153.5	156.4	139	133	21568	20975	−611	−93	523	79
0.5	7.2	153.5	157.1	139	135	21586	21353	−233	−36	377	56
0.6	7.2	153.5	157.8	139	136	21586	21760	174	22	235	32
0.7	7.2	153.5	158.5	139	138	21586	22150	564	78	69	10

FIGURE 2
DONOR REAL COSTS AND INCREASE IN CONSUMER EXPENDITURES FOR PROVIDING FOOD AID IN WHEAT

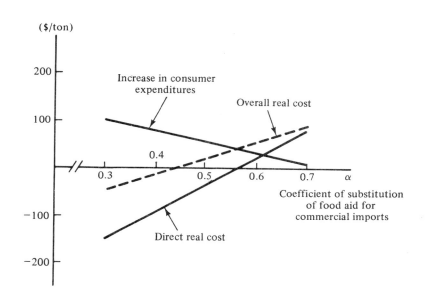

consumers, they will incur higher costs for their consumption of wheat. Assuming an inelastic demand for wheat in donor countries, the increase in consumer expenditures, expressed on a per unit of food aid provided, amounts to:

$$CC_t = \frac{D_t (P_t - P'_t)}{F_t}$$

where D_t = domestic utilisation of wheat in year t.

The results are also shown in Table 2. As it should be expected, the provision of food aid represents a higher cost to the consumers for any value of the coefficient of substitution. The lower the value of this coefficient, the higher the cost to consumers, as in this case the price effect of food aid is the highest. It should also be noted that higher consumer expenditures as a result of food aid are not restricted to consumers in donor countries alone. To the extent that increases in world prices are reflected in domestic prices of a country, consumers in that country will be equally affected. However, these distributional effects of food aid have not been considered in any detail in this chapter.

NOTES

1. The FAO Study noted that, in addition to export earnings, the level of foreign exchange reserves, foreign capital inflows and debt servicing should be taken into account in assessing the foreign exchange that a government allocated to the importation of cereals. But due to lack of complete data these factors could not be taken into account.
2. The results of the FAO study also indicate that out of the total increase in the effective import demand gap between 1979 (average of 1978–80) and 1985, commercial imports would meet 60 to 65 per cent and food aid the remainder.

REFERENCES

Abbott, P., 1979, 'Modelling International Grain Trade with Government Controlled Markets', *American Journal of Agricultural Economics*, Vol.61, No.1 pp.22–31.

Braun, J. von and B. Huddleston, 1988, 'Implications of Food Aid to Price Policy in Recipient Countries' in Mellor, J.W. and R. Ahmed, (eds.), 'Agricultural Price Policy for Developing Countries', Ithaca, NY: Cornell University Press.

Clay, E.J. and H.W. Singer, 1985, 'Food Aid and Development: Issues and Evidence: A Survey of the Literature since 1977 on the Role and Impact of Food Aid in Developing Countries' Rome: World Food Programme.

Food and Agriculture Organisation, 1983, 'Assessing Food Aid Requirements – A Revised Approach', *Economic and Social Development Paper* No.39, Rome.

Food and Agriculture Organisation, 1980, 'Principles of Surplus Disposal and Consultative Obligations of Member Nations', Rome.

Hopkins, R.F., 1983, 'Food Aid and Development: Evolution of Regime Principles and Practices', in *Report of the WFP – Government of the Netherlands Seminar on Food Aid*, Rome: World food Programme.

Huddleston, B., 1984, 'Closing the Cereals Gap with Trade and Food Aid', *Research Report* No.43, Washington DC: International Food Policy Research Institute.

International Wheat Council, 1984, 'Report on Shipments by Members of the Convention in 1982/83' prepared for the Food Aid Committee, London.

IDS and Africa Bureau Cologne, 1982, 'An Evaluation of the EEC Food Aid Programme', Report prepared for the Evaluation Service of the Commission of the EC, Institute of Development Studies, University of Sussex.

Konandreas, P., 1987, 'Responsiveness of Food Aid in Cereals to Fluctuations in Supply in Donor and Recipient Countries', in M. Bellamy, and B. Greenshields (eds.), *Agriculture and Economic Instability*, International Association of Agricultural Economists.

Maxwell, S.J. and H.W. Singer, 1979, 'Food Aid to Developing Countries: A Survey', *World Development*, Vol.7, No.3.

Minear, L., 1983, 'Development through Food: Some Non-governmental Reflections' in *Report of the WFP – Government of the Netherlands Seminar on Food Aid*, Rome: WFP.

Pinstrup-Andersen, P. and L.G. Tweeten, 1970, 'The Impact of Food Aid on Commercial Food Export', in *Proceedings of the Fourteenth International Conference of Agricultural Economists*.

Singer, H.W., 1987, 'Food Aid: Development Tool or Obstacle to Development?'

Development Policy Review, Vol.5, No.4.
Singer, H.W. and M. Diab, 1988, 'Food Aid: is it a Second Best Substitute for Financial Aid?', Institute of Development Studies, University of Sussex (mimeo).
Singer, H.W., J. Wood and T. Jennings, 1987, *Food Aid: The Challenge and the Opportunity*, Oxford: Clarendon Press.
Schuh, G.E., 1982, *Food Aid as a Component of General Economic and Development Policy: The Developmental Effectiveness of Food Aid in Africa*, New York: Agricultural Development Council.
Schultz, T.W., 1960, 'Value of U.S. Farm Surpluses to Underdeveloped Countries', *Journal of Farm Economics*, Vol.42, No.5.
Stevens, C., 1979, *Food Aid and the Developing World – Four African Case Studies*, London: Croom Helm.
US Agency for International Development, 1982, 'Sri Lanka: The Impact of PL480 Title I Food Assistance', *Impact Evaluation* No.39, Washington, DC.
World Food Programme/Government of the Netherlands, 1983, Report on Seminar on Food Aid, Rome.

3

The Disincentive Effect of Food Aid: A Pragmatic Approach

SIMON MAXWELL

INTRODUCTION

The 'disincentive effect' has been described by Baribeau and Gerrard [*1984*] as the 'storm-centre' of the debate on food aid. There is a voluminous literature of spiralling complexity. Yet a review of this literature shows it to be inadequate in many respects: incomplete with regard to its coverage of food aid commodities and uses; insufficiently well-rooted in the general literature on food policy; and biased to *ex-post* analysis, especially in countries with good time-series data of prices, production and imports. All this makes it of little practical help to donors and recipients faced with the day-to-day problems of planning, monitoring and managing a variety of food aid activities, in circumstances where data is poor and policy is changing fast. A new approach is needed.

This chapter sets out a pragmatic approach to the disincentive issue. It begins with a review of the existing literature, before outlining a new methodology. The methodology is then applied to three practical cases in Senegal, Ethiopia and Sudan. The main conclusions of the paper are that a systematic approach to disincentive analysis is feasible even in difficult cases; and that focusing on the link between food aid and food policy can help to ensure that food aid has positive, incentive, effects.

The literature on the disincentive effect was initiated by Schultz [*1960*] with the observation that if food aid were sold on the market, it would depress prices and lead to a loss of output. In the succeeding quarter century, the basic concept has developed and a good deal of empirical testing has taken place. A major review of the literature was carried out in 1979 by Maxwell and Singer.[1] This showed that thinking had divided into three streams, with the original focus on prices being supplemented by attention to agricultural policies and to labour markets. There was a subsidiary concern with the impact of food aid on food habits and a shift in taste away from locally produced commodities. The price disincentive was the dominant issue, with empirical testing having evolved from simple time-series analysis to multiple equation econometric models incorporating differentiated markets. There was a strong bias to India, with 12 out of 21 studies reviewed dealing with the Indian experience of PL480; of the remainder, only one dealt with a country in sub-Saharan Africa. The general conclusion of the empirical work was that a price disincentive had mostly been avoided by an appropriate mix of policy tools, including demand expansion, price support to producers and differentiated markets to increase consumption.[2]

The emphasis on policy in the price disincentive discussion illustrated the artificiality of separating price and policy discussions. Nevertheless, a separate strand of the literature dealt with the impact of food aid on government investment in agriculture; the administration of grain marketing; technical assistance to producers; attitudes to land reform; and the terms of trade between agriculture and industry. Again, the bias of empirical work was to India, but the general conclusion seemed to be that food aid was often 'sucked in by poor agricultural policy rather than itself being responsible for the vacuum which it fills' [*Maxwell and Singer, 1979: 225*].[3]

The third strand of the literature dealt briefly with labour disincentives, particularly the danger that food for work programmes might attract agricultural labour away from normal employment. There was little empirical work on this issue, although Stevens [*1979*] had concluded that in four African countries the effect did not occur.

Finally, with regard to food habits, Maxwell and Singer noted that increased familiarity with exotic products supplied as food aid could increase demand for non-indigenous food and thus lead to greater dependency in the long run. Wheat and dairy products were the main culprits.

Research since 1979 has contributed to disincentive concepts, measurement techniques and policy findings. As regards *concepts*, a recurrent theme of the literature has been that the potential for disincentives is in practice restricted by the way in which food aid is used. In the first place, much cereal food aid simply replaces commercial imports, at least in non-emergency situations: it cannot be held responsible for disincentive effects that would have occurred in any case as a result of commercial imports.[4] In the second place, the literature has stressed that disincentives can be avoided if food aid is associated with additional consumption, so that the depressing effect on prices of the additional supply of food is matched by the stimulating effect of extra demand. Additional consumption can be stimulated directly, by distributing food to hungry people; or indirectly, by increasing expenditure on poverty alleviation programmes.[5] It is argued that a combination of measures along these lines can significantly reduce the risk of disincentives – although not the need to monitor for them and plan remedial action in case they do develop.[6]

Despite this focus on the circumstances in which disincentives may be avoided, recent research has continued to add new layers of complexity to disincentive analysis. There has been particular concern with the relationship between food aid, food policy and overall development strategy.[7] For example, the United States Agency for International Development [*1983*] has taken material concerned with the relationship between food aid and local market structures and related it to thinking on linkages between agriculture and other sectors in the development process [*Mellor, 1976, 1980*]: it argues that beneficial linkages can be disrupted by food aid which causes an 'uncoupling' of processing industries from the agricultural sector. Backward and forward linkages are lost. Buchanan-Smith [*1988*] is also concerned with market structures. She argues that food aid can undermine the incentives to local traders and cause greater year to year variability in prices by reducing the level of inter-annual storage.

Of course, the debate about these and earlier forms of disincentive needs to be kept in perspective. In a valuable guide to addressing Bellmon amendment concerns on the disincentive effects of food aid, USAID [*1985*] makes the important point that disincentive relationships are progressive. Thus, it is suggested that a disincentive is only likely to occur if PL480 supplies amount to over two per cent of the total staple consumption and that a full analysis is needed only when the figure rises to ten per cent.

There are two points that are not yet reflected in the literature. The first is the link between the disincentive and the dependency debates (see Maxwell and Singer [*1979*]). The latter has been particularly concerned with foreign exchange dependence caused by food aid supplies and fiscal dependence caused by food aid sales. It is arguable that these constitute a disincentive to self-reliance and should be examined as potential negative effects of food aid [cf. *Stewart, 1986*]. A similar point is made by those who argue that if food aid provides balance of payments support, it can buttress an overvalued exchange rate and lead to implicit taxation of agriculture [*CIDA, 1984*]. However, all these arguments could also be applied to financial aid.

THE DISINCENTIVE EFFECT OF FOOD AID

The second point arises from the increasingly close connection between food aid policy and food policy generally, particularly in so far as the latter is concerned with setting normative standards for prices or other instruments of agricultural policy [*FAO, 1984; WFC, 1985a, b*]. The price disincentive has normally been interpreted as meaning that prices are below what they would otherwise be. However, if food aid dampens inflationary pressure it may undermine scarcity prices so that these are lower than they would otherwise be, but about what they 'ought' to be: perhaps disincentives should be measured not in terms of lower prices, but in terms of 'deviations from optimal prices'.[8] Of course, this formulation begs the question of what is the optimal price; it is also a concept that might be difficult to apply to policy disincentives, where optimality is even harder to define. Two of the three case studies below (on Sudan and Senegal) explore further this idea of relating disincentives to a desired food policy frame.

Turning to questions of *measurement*, there is a traditional dichotomy in the literature between attempts to measure the disincentive effect using econometric models and more pragmatic approaches, using what Clay and Singer [*1985*] have described as a 'checklist approach'. The former are mostly to be found in academic studies (see, for example, Blandford and Plocki [*1977*]); the latter, as Clay and Singer point out, are characteristically found in evaluation studies by food aid donors. The non-formal approach is seen as being less rigorous, particularly in dealing with inter-sectoral linkages and dynamic growth effects, but conceptually easier and much less demanding of data.

Recent work on the measurement of disincentive effects has derived from both formal and non-formal approaches. With regard to formal methods, Bezuneh and Deaton [*1981*] have produced a multi-equation model which gives greater weight than earlier work to positive income effects; and Abbott and McCarthy [*1982*] have approached the analysis of multiple-objective food aid with a welfare analysis. Bezuneh *et al.* [*1988*] have applied household modelling techniques to the analysis of food for work in Kenya. These all belong to the tradition of partial equilibrium analysis in food aid studies. Clay and Singer have pointed to a new stream of literature which is concerned with general equilibrium models. The value of these is that variables otherwise treated as exogenous may become endogenous. The kind of result obtained is exemplified by Ahluwalia's [*1979*] study of India which concludes that 'a relative decline in the price of food grains (*prima facie* evidence of a disincentive effect) is compatible with an overall marginally positive impact on agricultural output ...' (quoted in Clay and Singer [*1985: 36–7*]). Cathie provides a comparable example from Botswana in this volume.

On the non-formal side, advances have been made in developing checklists of information required. Many of these are related to general food policy analysis, but an exception is the Bellmon guide referred to earlier [*USAID, 1985*]. This is notable for its careful analysis of agricultural strategy and its combination of quantitative and qualitative analysis. Nelson [*1983*] provides a case study of Bangladesh using similar principles.

The literature is notable for the prevalence of longitudinal case-studies and for the absence of cross-section studies covering a number of countries. This might be an interesting area for future econometric work. Nevertheless, it does not seem that methodological improvements are likely to produce formal models of great usefulness in the near future, particularly in low income food deficit countries where time series data is poor and where structural changes associated with adjustment policies make it difficult to use models based on past data for future purposes. The modellers themselves are not optimistic: Blandford and Plocki [*1977: 45*] conclude that 'considerable dangers exist in the use of these, or any other estimates, for policy purposes'.

Finally, the *findings* of disincentive analysis. As noted above, the established view in the literature is that food aid has the capacity to cause disincentive effects, but that these can be and often are avoided by government policy. This is a view confirmed in recent evaluations [*Maxwell (ed.), 1983; USAID, 1983; CIDA, 1985*] as well as in recent literature reviews [e.g. *Baribeau and Gerrard, 1984; Clay and Singer, 1985; Raikes, 1988; Thomas et al., 1989*], although Jackson [*1982*] has provided evidence of labour disincentives on particular projects. Clay and Singer conclude [*41–2*] that 'the debate on the past macro-economic and agricultural impact of food aid remains inconclusive ... [however] massive disincentive effects do not seem to have occurred'.

To the extent that disincentive effects have been measured in econometric models, the effect is generally presented as being linear. The size of the effect varies substantially. Blandford and Plocki [*1977*] suggested that an extra one million tons of cereal food aid to India led to a decline of the cereal price index in the same year of 2.5 units with a negative effect on production that stabilised after seven years at 149,000 tons p.a. On the other hand, Bezuneh and Deaton [*1981*] found a negligible impact on production in Tunisia; and, for Brazil, Hall [*1980*] found that long-run multipliers were positive, so that food aid actually led to increased production.

Recent research has focused on sub-Saharan Africa (SSA) and

confirms this mixed picture. A recent review concluded that 'among the major recipients of food aid in sub-Saharan Africa there is evidence of market disruption and policy disincentive effects' and cited evidence from Somalia, Sudan, Tanzania and Botswana. However, it found that 'experience in some other countries receiving smaller quantities of food aid has been evaluated more positively': studies of Rwanda, Kenya, Cameroon and Lesotho are cited in support of this conclusion [*Thomas et al., 1989: 17–19*].[9] In Ethiopia, the second largest recipient in SSA after Sudan, there is no evidence of major disincentives, although, as discussed in the case study below, 'warning signals' have been flashing for a price disincentive [*ibid: 19–20*]. Reflecting the conceptual shift identified earlier in the article, the review notes that 'increasingly, donor and recipient attention has been focused on linking food aid to a major restructuring of domestic food markets, often as part of an economy-wide structural adjustment'.

The development of the literature since 1979 has helped to broaden the geographical spread of disincentive analysis and strengthen the conceptual framework. Nevertheless, weaknesses remain and five of these are worth discussion. First, the literature is incomplete as regards the range of commodities and the uses to which food aid is put. Characteristically, the disincentive literature is concerned with US wheat, provided as 'programme' food aid for sale on recipient markets. In practice, food aid is far more diversified in commodities, sources and uses than the thrust of the literature would suggest. Over a quarter of food aid by value consists of non-cereal products (mostly milk powder and butter oil) and only a little over half is provided for sale: the rest is for emergency distribution or for use on food for work or supplementary feeding projects.[10] The US now accounts for only 60 per cent of total cereal food aid and most recipients receive food aid from several donors. A framework of analysis is needed which allows for multiple donors, a range of commodities and a variety of uses.

The second weakness of the literature has to do with the treatment of different types of disincentive. The price disincentive has been well specified and well studied but the policy disincentive has not. Labour disincentives are barely treated and scant attention is given to the relationship between food aid, food habits and potential disincentives. Most important, few studies look simultaneously at all these aspects. A new approach should provide the opportunity to deal systematically with disincentives across the board.

The third weakness relates to levels of analysis. The bulk of the literature looks at the national level with only passing recognition of the fact that food aid may have a disproportionate impact at regional or local level. This bias is redressed in the literature dealing with the labour disincentives of food for work; but needs to be tackled more generally.

The fourth area of weakness concerns the nature of disincentive analysis, which characteristically remains *ex-post*, linear, data-hungry and biased to countries with above-average food systems. Perhaps the most damaging defect of existing methods is the emphasis on *ex-post* analysis, often using very long time series of doubtful reliability. An improved methodology should enable decisions to be made at the margin, based on concurrent analysis of the evolving food situation.

Finally, it is apparent that the food aid literature has lagged behind the more general food policy literature in relating food indicators to some desired food policy frame. The precise content of food policy (for example, the balance to be struck between food crops and export crops)[11] is likely to be controversial in most countries and there is no blueprint to be followed. Nevertheless, the food aid literature is beginning to recognise the benefits to be derived from the analytical methods of food policy analysis.

The requirements for a new approach can now be defined. It should be comprehensive as regards commodities, uses and types of disincentive effect; range across levels of analysis; and provide food aid planners with a tool that can be used not only for *ex-post* evaluation but also for planning and managing new and ongoing food aid activities. Most important, it should set individual food aid programmes in the wider context of food policy. This is not an impossible task, as we shall see.

A PRAGMATIC APPROACH

A practical approach to identifying the potential disincentive effects in situations where data is poor and quick responses are needed is set out in three stages. First, the theoretical literature on food aid is summarised in the form of 19 'scenarios', statements about the different ways in which disincentive effects might appear. Second an 'early warning' indicator is suggested for each scenario, to provide a practical tool for day to day monitoring of disincentives. And third, some suggestions are made about the more detailed analysis that is necessary if a causal link is to be established between food aid deliveries and apparent disincentives.

THE DISINCENTIVE EFFECT OF FOOD AID

Definitions and Taxonomy: 'Disincentive Scenarios'

As the literature shows, the term 'disincentive' is used in different ways. It can refer to a negative impact on (i) the production of the particular commodity being supplied as food aid; or (ii) food production more narrowly; or (iii) agricultural production as a whole.[12] In some recent literature, the term 'disincentive effect' has been taken to refer not just to agricultural production but to food security more widely, including both production and consumption aspects. Furthermore, the negative impact may either be localised at the site of a particular food aid project ('micro-disincentive') or felt on a wider scale at the regional or national level ('macro-disincentive'). Finally, any disincentive may be short term, lasting only as long as the food aid project; or long term, reflecting permanent damage to agricultural production. In principle it seems right to take the most general view: disincentives should be avoided for any aspect of agricultural production or food security at any level of analysis and for any time period.

A 'disincentive effect' can take various forms. The four main types discussed in the literature act through: (i) prices, both in completely free markets or, as is more commonly the case, in imperfect markets with different degrees of government intervention; (ii) policies, in the sense that governments are led by food aid to neglect agriculture or other aspects of food security; (iii) food habits, where food aid is said to cause food preferences which lead to greater import dependence; and (iv) labour, where food aid may disrupt labour supply to agriculture.

Disincentives can manifest themselves in different ways and their diverse nature makes them difficult to model comprehensively.[13] They can, however, be presented in narrative form. A list of 19 disincentive scenarios is provided in Figure 1, working through prices, policies, food habits or labour at both macro and micro levels. The list is fundamental to the analysis of disincentives.

Scenario 1 is the classic 'Schultzian' disincentive whereby food aid is sold on the market, supply increases and prices fall. It is modified, however, to take account of the notion discussed earlier that there is an optimal price above which price falls do not matter but below which they may cause a disincentive. The free market implicit in this scenario is uncommon, so Scenarios 2 and 3 allow for the intervention of a public procurement agency. Scenario 4 allows for the possibility that food aid may cause a price disincentive even without being sold, if it contributes to the overhang of stocks; and Scenarios 5 and 6 allow for the possibility

FIGURE 1
FOOD AID DISINCENTIVES: SUMMARY OF SCENARIOS

Macro disincentives

1.1. Prices

1. Public sector food aid sales cause prices to fall below optimal levels.
2. The availability of food aid allows the Government procurement agency to pay 'low' prices to farmers.
3. The availability of food aid allows the Government procurement agency to buy less, which in turn leads to a price fall.
4. Food aid contributes to excessive stocks which overhang the market and depress prices below optimal levels.
5. The availability of food aid in one region displaces imports to that region from others, with a consequent downward effect on prices.
6. The availability of food aid in a region enables exports to the rest of the country to rise, causing prices generally to fall below optimal levels.

1.2 Policies

7. Food aid causes a national policy disincentive, allowing government to neglect agriculture and/or food security.

1.3 Food habits

8. Food aid causes taste changes which reduce demand for local staples.

1.4 Labour

9. The availability of food aid discourages labour migration, pushing up wages and leading to a loss of output.

2. Micro disincentives

2.1 Prices

10. Prices fall below the optimal level because food aid recipients sell a part on the market.
11. Prices on the local market fall because food aid recipients buy less.
12. Food prices fall because the availability of food aid allows a greater proportion of home production to be sold.
13. Food aid contributes to excessive local stocks in the public or private sectors, which overhang the market and drive prices below optimal levels.
14. Food aid discourages local storage which in turn increases price instability and reduces incentives.

2.2 Policies

15. The availability of food aid encourages government to neglect agriculture and/or long term food security.

2.3 Food habits

16. Food aid causes taste changes which reduce demand for local staples.

2.4 Labour

17. Competition for labour between food-aid supported activities and local agriculture reduces labour input to agriculture and leads to lower production.
18. Food aid creates a 'food dole' and reduces the incentive to work.
19. Wages are bid up because of competition from food aid-supported activities, with adverse effects on agricultural employment and output.

that food aid might disrupt inter-regional food trade. In all these cases, the indicator is not a price fall in isolation, but a fall below the 'right' price.

Scenario 7 is the classic policy disincentive, broadened to include not only agricultural production as a dependent variable but food security more generally. Implicit in this formulation is the notion that food aid should contribute to long term food security objectives. Scenarios 8 and 9 deal with food habits and labour markets, allowing for the possibility that food aid may distort food markets or labour migration patterns at the national level.

Scenarios 10–19 move from the national level to the project or 'micro' level, which may be a single village or as large as a region. Prices may be affected and may fall below the optimal level because food aid recipients sell part of what they receive (Scenario 10), buy less than they normally would (Scenario 11), or sell more than usual of their own production (Scenario 12). Stocks (Scenarios 13 and 14) are more likely to be held in private hands, often at household level. An important set of micro-disincentives works through labour, with food aid having the potential to compete with local agriculture (Scenario 17), create a 'food dole' (Scenario 18) or force wages above reasonable levels (Scenario 19).

It is worth making the point that these scenarios are not all equally likely to occur and that different kinds of food aid tend to be associated with different kinds of disincentive risk. Thus programme food aid for sale through the public distribution system is more likely to have adverse effects acting through the first half of the list in Figure 1; whereas localised food for work is more likely to work through the mechanisms identified in the second half of the list.

Early Warning Indicators

The second step in the analysis is to identify an indicator for each scenario, so that planners and project managers will have early warning that a disincentive may be developing. The market price of basic staples might be one such indicator: if it remained above the guide price, there would be no need for concern over price disincentives; but if it fell below the guide price, so that a disincentive might occur, it would be necessary to ask whether food aid were responsible and what might be done.

Preparing a list of 'key indicators' presents problems that are both conceptual and practical. The conceptual problems derive from the underlying volatility of cereal and labour markets and from the lack of consensus on what constitutes an optimal policy for agricultural development and food security. In the case of cereal markets, for example, not only do prices vary markedly from place to place, from one part of the year to another, and from year to year; but it is right that they should do so, to reflect changing levels of supply and demand, remunerate traders who move food around the country and provide incentives for inter-seasonal and inter-annual storage (see Shuttleworth [1988]). A single indicator price would therefore be misleading, too low in some periods, too high in others, either concealing a potential disincentive or indicating a threat where in reality none existed. The solution may therefore be to provide a range of prices.

Where this range should be located is another conceptual issue. Neoclassical economic theory would suggest that the 'right' price, on average, should be close to international prices, adjusted for the costs of trade: this means that for importing countries or regions, the price would be set at the level of import parity; for exporting countries or regions, it would be set at the level of export parity [Timmer, 1986; World Bank, 1986b]. However, Lipton [1987] has criticised pricing policy based on these principles as having adverse distributional consequences and being very difficult to apply in isolation when there are many other distortions, such as an over-valued exchange rate, affecting the economy. In Figure 2 the starting point is the import parity price at the shadow rate of exchange; but, as the Ethiopia case study will show, this indicator cannot be applied mechanically.

Similar conceptual problems arise in the case of indicators for a policy disincentive. Here the task is to specify indicators of a satisfactory food security policy which covers the production, marketing and consumption of food. If a food strategy with specific targets has been prepared, this task may be straightforward; otherwise, indicators will have to be defined. The literature reviewed earlier suggests that these will include government expenditure on agriculture, intersectoral terms of trade and the establishment of specific food security interventions such as public works and targeted nutrition programmes.[14]

Finally, there are conceptual problems in defining an indicator for labour disincentives. If the labour market is disrupted by food aid, the effect should show up in wage rates. However, these also vary seasonally and inter-annually, in both nominal and real terms, being higher during the harvest period than in the dry season and higher in good years than in bad. There is also the problem that what might look like a disincentive (higher wages) is a desirable factor for agricultural labour. The historical trend of real wages is suggested as an indicator

FIGURE 2
FOOD AID DISINCENTIVES: SUMMARY OF KEY INDICATORS

SCENARIO NO. (from Figure 1)	SUMMARY	KEY INDICATOR	REFERENCE POINT AND LIMIT OF ACCEPTABILITY	
1	Food aid sales	Market prices	Import parity	±15%
2	Low proc. prices	Procurement prices	Import parity	±15%
3	Low gov't purchases	Market prices	Import parity	±15%
4	Excessive stocks	Market prices	Import parity	±15%
5	Regional purchases down	Market prices	Import parity	±15%
6	Regional exports up	Market prices	Import parity	±15%
7	Poor food security policy	Food policy	Plan targets	−20%
8	Taste changes	Market prices	Import parity	±15%
9	Migrant labour falls	Wage rates	Real historical	±15%
10	Recipients sell food	Local market prices	Import parity	±15%
11	Recipients buy less	Local market prices	Import parity	±15%
12	Recipients sell produce	Local market prices	Import parity	±15%
13	Excessive local stocks	Local market prices	Import parity	±15%
14	Greater price instability	Local market prices	Import parity	±15%
15	Poor local food security	Food policy	Plan targets	−20%
16	Taste changes	Local market prices	Import parity	±15%
17	Competition for labour	Wage rates	Real historical	±15%
18	Food dole	Wage rates	Real historical	±15%
19	Wages driven up	Wage rates	Real historical	±15%

in Figure 2. However, such a trend will clearly require careful interpretation.

Turning to practical issues, the key difficulty lies in finding recent and accurate statistics. The availability of data on aggregate food aid flows is a major problem: in Senegal, for example, a comparison in 1986 of data series produced by five different organisations showed discrepancies of up to 100 per cent in individual years.[15] Similarly, detailed price data may be unavailable or unreliable: again, in Senegal in 1986, the official retail price index did not include the price of the basic staple, millet.[16] In response to problems of data availability, the procedure outlined below focuses on the smallest possible number of indicators, selecting those most likely to be available. This has the added advantage that first stage disincentive monitoring can be based on a few simple indicators.

Putting these principles into practice produces the summary of key indicators in Figure 2, one for each of the 19 scenarios listed in Figure 1.

A reference point and limit of acceptability are also given for each indicator. The number of indicators has been kept to five, three of which have to do with food prices, one with wage rates and one with overall movement in food policy. The reference points are as described above, with limits of acceptability that are essentially arbitrary. It should be stressed that this table is highly speculative and that it will need to be adapted to particular country or project situations. Nevertheless, it does provide a basis for initial monitoring, as will be seen in the case studies below.

Causal Relationships

If the key indicators suggest a potential disincentive, the question arises whether food aid is responsible. Causality is often hard to establish. For example, a fall in prices may result from a combination of government policy, the size of the harvest, traders' expectations and, perhaps, the availability of food aid. How are food aid managers to know whether food aid is contributing to disincentives?

The first and, in the end, the only definitive answer to this question may be found in the scientific literature on food aid, where causality is established by constructing testable hypotheses and applying statistical tests. As noted above, the statistical techniques available are becoming increasingly sophisticated and in many cases require implausibly large amounts of data. Nevertheless, simple regressions or correlations may be feasible and illuminating. For example, simple regression analysis was used in Ethiopia to explore the relationship between additional food aid and commercial food imports [*FAO, 1982; Maxwell, 1986c*].

If statistical testing proves not to be possible, the second answer to the problem is to use less formal methods. These may still be based on quantitative data but will rely not on formal modelling but rather on a systematic review of the evidence and on the analyst's careful judgement as to the balance of probability that specific hypotheses about the role of food aid are correct. Non-formal but resolutely rigorous analysis is characteristic of the checklist approaches reviewed above. It is the method used of necessity in the case studies that follow. Rapid rural appraisal will have a part to play in non-formal disincentive analysis and such an approach has the advantage of allowing greater beneficiary participation in monitoring and evaluation.[17]

DISINCENTIVE ANALYSIS IN PRACTICE: THREE CASE STUDIES

The method described in the previous section is intended to apply across countries, including those where food systems are rudimentary, data is poor, government policy is uncertain and variable and food aid is being received from a variety of donors for a variety of purposes. The three brief case studies in this section draw on more detailed country reviews for Ethiopia and Senegal that were prepared at an earlier stage of research [*Maxwell, 1986b, 1986c*]; and on more recent work on food policy issues in the Sudan [*IDS, 1988; Sudan, 1988*]. The case study from Ethiopia deals with local price disincentives; that in Senegal with macro-policy disincentives; and that in Sudan with changes in taste. The focus is on the value of the early-warning indicators and the systematic analysis of potential disincentive scenarios.

Ethiopia, Senegal and Sudan meet most of the requirements for a difficult test of the disincentive methodology, but otherwise provide contrasting cases. Thus, Senegal is on average only 50 per cent self-sufficient in cereals, with large imports of rice and wheat, mostly, but not exclusively, for urban consumption; food aid has typically provided less than a quarter of total imports. Ethiopia is much closer to total cereal food self-sufficiency, averaging 95 per cent before the drought in 1984/85; however, food aid provides over 80 per cent of total cereal imports. Sudan has a large exportable surplus of sorghum in most years, but wheat accounts for about 20 per cent of total cereal consumption and over 80 per cent of this is provided by aid.[18]

Price Disincentive: Ethiopia

The first case study is concerned with possible price disincentives immediately following the 1984/85 drought in Wollo region. Wollo received a substantial share of the 1.2 million tons of cereals shipped to Ethiopia in the emergency year, 1984–85, and continued to receive food aid for both development purposes and emergency relief in the following year. Relief needs for a total of 1.9 million destitute people in Wollo were estimated at over 300,000 tons of food in 1986, and in addition 23,000 tons were to be provided for developmental food for work. At the same time, however, food production recovered dramatically in the growing season of 1985, with national cereal and pulse output estimated to have increased from 4.3 million tons in 1984/85 to six million tons in 1985/86, only slightly below trend. The combination of a large cereal harvest and very large food aid distribution posed a clear risk of local price disincentives, corresponding to scenarios 10–13 in Figure 1.[19]

Figure 3 provides the local price data required for a disincentive indicator. The data cover sorghum prices in the Woldia market for a normal period (1981/82–1983/84) and for the period January 1984–

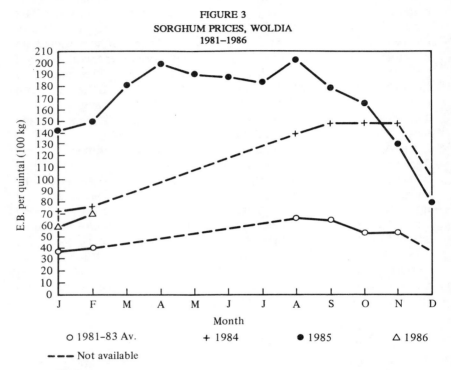

FIGURE 3
SORGHUM PRICES, WOLDIA
1981–1986

○ 1981–83 Av.   + 1984   ● 1985   △ 1986
– – – Not available

Source: RRC

February 1986.[20] All these series are incomplete but they show a generally rising trend, superimposed on a seasonal movement which takes prices to a peak between July and September, before the beginning of the harvest in October. The very marked increase in prices during 1984 and the first half of 1985 is notable; also the extremely rapid decline in prices from the peak in August 1985 of over EB 200 per quintal (100 kg) to the trough in December 1985 and January 1986 of below EB 60 per quintal (US$1 = EB 2.07). This is still above the normal price for the time of the year but is very much below the price for the previous year and does not allow for inflation over the five year period since 1981/82. It could therefore be considered *prima facie* evidence of a price disincentive effect.

Figure 4 provides additional information. Superimposed on the original chart are the Agricultural Marketing Corporation purchase price for 1985/86 of EB 23–27 per quintal and the 1985 import parity

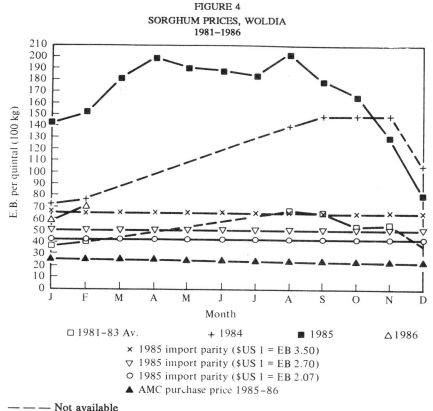

price for Wollo, calculated for a range of shadow exchange rates. It can be seen that the free market price at its lowest was still more than twice the AMC purchase price;[21] however it had entered the range of import parity prices, being below the highest import parity but above the lowest. It seems reasonable to conclude that the price of sorghum in the Woldia market had declined extremely fast and had reached levels where it might be beginning to fall below the right price as represented by import parity. In this sense, the warning lights were flashing.

The urgent question facing food aid managers was whether food aid was responsible for the price fall and if so what should be done about it. In theory, it would have been possible to construct a model relating food prices to production, marketing behaviour and alternative uses of food aid, but none of the required data were available. It was therefore necessary to have recourse to non-formal methods. There was strong circumstantial evidence from talking to key informants in Wollo that very large distributions of food aid, particularly as emergency relief, contributed to the fall in grain prices in late 1985. Over the country as a whole, prices appeared to have fallen fastest in areas where food aid distribution was largest [*USAID, personal communication*]. This of course did not mean that emergency relief was not required in Wollo during the period: clearly, many destitute families needed help. However, relief in kind using imported food would not be appropriate if it were to have negative effects on local markets. The solution, therefore, was to support simultaneously both local markets and the food entitlements of poor families by using aid money either to buy food locally for free distribution or to fund cash for work. With local markets being volatile, there would have to be flexibility between food and cash. However, a graduated shift from food to cash did seem the best way to avoid disincentives [*Maxwell, 1986c: 24–6, 36*].

Policy Disincentive: Senegal

The second case study is concerned with the possible relationship in Senegal between increasing food aid on the one hand and, on the other hand, poor food and agricultural policy leading to low growth, increasing import dependence and very high levels of malnutrition: Scenario 7 in Figure 1.

The facts are easily stated. On the one hand, cereal food aid tripled between the early 1970s and the early 1980s, averaging over 50,000 tons a year in the latter period and accounting for around 5 per cent of consumption. On the other hand, the agricultural growth rate was low at only 0.3 per cent per annum (1973–83), national food self-sufficiency declined to under 50 per cent in the 1980s and malnutrition rates were high. Furthermore, and corresponding to the key indicator in Figure 2, poor government policy had played an important part in the failure of the food sector, along with deteriorating terms of trade and recurrent drought. For example, income distribution was poor, agriculture accounted for only 14 per cent of government investment in the 1970's and the sale of imported cereals at subsidised prices might have affected the market for local cereals.[22] The question for food aid managers was whether food aid was in some way responsible for this poor performance; and if so what remedial action might be taken.

These questions were addressed in late 1985 [*Maxwell, 1986b*], testing food policy in Senegal against the indicators discussed above. The main finding was that improvements had been made on the production side, so that the policy framework was conducive to faster and more equitable growth; but that adjustment measures taken since 1979 had undermined the food entitlements of the poor and created a crisis of consumption.

The economic crisis had been tackled by a series of adjustment programmes, notably the Plan à Moyen Terme de Redressement Economique et Financier (PREF) of 1979. In 1984, this had been complemented by a new agricultural policy, the Nouvelle Politique Agricole (NPA) [*Sénégal, 1984*]. The main orientations of the policy were summarised in the press as 'moins et mieux' (less and better). The document noted that the rural development parastatals had become authoritarian (p. 36) and directed that they should become 'souple et léger' (flexible and light of touch). Markets were to be liberalised. The policy document did not set out precise targets for the share of investment devoted to agriculture. It noted, however, that government support to agriculture had risen by an average of seven per cent per annum between 1977/78 and 1982/83. In output terms, the new policy aimed at an improvement in cereal self sufficiency from only 50 per cent in the early 1980s to 75 per cent by the year 2000.

Following the publication of the NPA, a free market was introduced for millet, the main domestic cereal, subject to a government-supported minimum price. In the case of rice, the government continued to control imports, but the retail price of rice more than doubled and by 1985 rice was being sold above world market prices, generating a profit for the government of 20–25 per cent. In the case of wheat, imports were again controlled and the price of flour was increased by approximately 25 per cent in October 1985, turning a subsidy of approximately ten per cent into a profit of almost the same percentage.

All this led to a favourable conclusion about the then role of food aid in Senegal as far as production was concerned. Under the policy regime prevailing in Senegal, food aid would not be held responsible for major disincentive effects. This was true even though food aid was associated with a markedly import dependent food system.[23]

However, the policy framework was seen as containing a major flaw in that it failed to take account of the impact on consumption of cuts in government expenditure and increases in the price of basic staples. A case was made for targeted food or nutrition interventions funded from food aid counterpart funds. In its simplest form, this could be done by using part of the resources of the common counterpart fund which already existed. On a larger scale, it would be possible to think of expanding cereal shipments to Senegal with the dual purpose of providing balance of payments relief and funding targeted nutrition interventions. A policy recommendation for food aid donors was that they should provide as much wheat and rice as possible to Senegal, ensure that this was sold at a premium price and devote the resources to supporting the production and consumption of local staples [*Maxwell, 1986b: 21ff*].

There is a postscript to the analysis of food policy in Senegal, in that riots in the spring of 1988 led the government to reduce the price of staple cereals. This led to the cancellation of a programme initiated by France to use food aid counterpart funds to support the processing of local millet because imported rice became cheaper than processed millet [*Jost and Gabes, 1988*]. This seems to indicate the emergence of a disincentive, although not necessarily one caused by food aid. Clearly, monitoring needs to be continuous.

Change in Food Habits: Sudan

The final case study deals with the possible impact of food aid on food habits in Sudan: Scenario 8 in Figure 1. The country shows a clear trend to increasing consumption of wheat, the share of which rose from 10 per cent of cereal consumption in 1970/71 to over 20 per cent by 1986/87. At the same time, total imports rose from about the same level as domestic production to something like four times domestic production, so that by the end of the period, the country was over 80 per cent dependent on imports. Food aid provided some 85 per cent of imports.[24] The scenario, then, is one in which food aid is associated with increasing consumption of an exotic commodity, with the clear potential for disincentives to local cereals.

In order to trace whether food aid is responsible for increasing import dependency, Figure 2 identifies the price of wheat as the key indicator, with the import parity price as the reference point. If wheat is sold at the import parity price, or even higher, as was the case in Senegal in 1985, taste changes and the resulting shift in consumption might occur even without food aid.[25] If, however, food aid wheat is sold below the import parity price taste changes are more likely to occur.

In Sudan, wheat and flour distribution are controlled by the government and bread is sold at a subsidised price. The total budgetary cost of the subsidy in 1987/88 was nominally LS (Sudanese Pounds) 262m (US$1 = LS 2.50), equivalent to 7 per cent of revenue.[26] However, the fact that most imported wheat was in the form of free or subsidised food aid protected the budget from the cost of the subsidy: IDS [*1988: 79*] calculated that the government actually made a profit of some LS 80m per annum on its wheat account.

Whatever the budgetary implications of the wheat subsidy, the net effect of providing cheap bread has been to change the relative prices of basic staples in Sudan. Comparing the price of bread with the price of sorghum, IDS [*ibid: 83*] calculated that the real price of bread fell by 60 per cent between 1970/71 and 1984/85: more than 30 per cent of the observed increase in the per capita consumption of bread during the period could be explained by this decline in the relative price of bread. Thus, a substantial part of the increased dependency on wheat imports can be explained by the food aid financed cheap bread policy. Without food aid the policy would not be sustainable.

Of course, this is not a necessary outcome of providing wheat aid. As the IDS notes, 'the underlying problem may not be consumption of wheat *per se*, but the fact that it is subsidised, with consequent losses to producers and the government budget' [*1988: 85–6*]. Options for the government include raising the price of bread to 'economic' levels, raising other taxes to cross-subsidise cheap bread, replacing the generalised subsidy with a targeted subsidy, or substituting sorghum for wheat in bread. A combination of these is likely to be pursued, with the active support of donors anxious to avoid further disincentives.

Conclusion

The three case studies illustrate food aid at work in very different circumstances; yet the conclusions are strikingly similar. It is clear in all three cases that sophisticated econometric analysis is not feasible: data are either unavailable or unreliable; the policy environment is unstable; and food aid programmes themselves are heterogeneous. Furthermore, econometric analysis would be unlikely to give the guidance on fine tuning that is necessary to improve the management of specific food aid programmes. However, the difficulty of econometric analysis does not provide an excuse for lack of rigour. An important lesson to be drawn from the case studies is that a comprehensive taxonomy of disincentive effects makes it possible to carry out a systematic analysis even where data are poor. The case studies can be said to have validated the list of scenarios in Figure 1, especially the distinction between micro and macro levels of analysis and the separate consideration of price, policy, labour and food habit disincentives. They also illustrate the value of a step-by-step procedure building on the notion of early warning indicators and clear reference points.

If the methodology provides an adequate basis for investigating disincentives, what substantive lessons can be drawn about disincentives from these cases? In all three, the warning lights were flashing to indicate the possibility of disincentive effects. In Ethiopia, the main focus was on the low producer price of cereal staples; in Senegal, it was on the neglect of consumption issues in food policy; and in Sudan it was on the impact of cheap bread on import dependence. In all three cases, policy changes were needed but a constructive role remained for food aid.

These conclusions are broadly in line with the literature on food aid. The possibility of disincentive effects is modified by government policy, which is not normally determined primarily by the flow of food aid. However, and this is a finding not commonly stressed in the literature, the effects of food aid can be disproportionate at subnational level. The findings in Ethiopia, particularly, underline the importance of careful monitoring of food interventions in isolated communities where food markets are volatile.

In general, the case studies support the view that food aid needs to be integrated into an overall food strategy if disincentives are to be avoided and the full potential of food aid incentives realised.

NOTES

The original research on which this chapter is based was financed by the World Food Programme and published in a series of IDS Discussion Papers as well as in guidelines issued by the WFP [*Maxwell, 1986a, 1986b, 1986c; WFP, 1987*]. It has since been extended through work on food policy in Sudan, financed by the European Community [*IDS, 1988; Maxwell, 1989*]. The results have been subject to critical review in various seminars and in meetings organised by the DSA (UK Development Studies Association) and EADI. The views expressed are those of the author alone and do not in any way commit the WFP, the EC or the many colleagues and seminar participants who have made useful suggestions.

1. The literature was also reviewed by Isenman and Singer [*1977*]; more recent reviews include those by Baribeau and Gerrard [*1984*], Clay and Singer [*1985*] and Thomas *et al.* [*1989*].
2. The best-known papers on the price disincentive issue are those on Colombia by Dudley and Sandilands [*1975*]; and on India by Mann [*1967*] and Rogers *et al.*

[1972].
3. On the policy disincentive, see particularly Isenman and Singer [1977], Barlow and Libbins [1969], and Merrill [1977].
4. Clay and Singer [1985: 16]; Stewart [1986]; Singer [1987], Fitzpatrick and Storey [1989]. The last-named make the important point that food aid which replaces commercial imports may still create disincentive effects at regional and local level if it is additional to local supplies. A parallel argument could be made about policy disincentives.
5. Stewart [1986], Fitzpatrick and Storey [1989]. It is important to draw attention to the multiplier effect of increased government expenditure on food demand, which was an important feature of earlier food aid analysis [e.g. FAO, 1961]. See also Maxwell and Singer [1979].
6. Fitzpatrick and Storey have argued that under certain circumstances a disincentive may actually be desirable, for example, because lower food prices are good for the poor [1989: 246].
7. Documented, for example, by Thomas et al. [1989].
8. This idea has a pedigree in the earlier food aid literature. See, for example, Dandekar [1965].
9. Lesotho is a surprising entry in the category of countries receiving small amounts of food aid. Deliveries in 1986–87 were 20 kg per capita, the sixth highest in SSA [Thomas et al., 1989: 3].
10. These figures are estimates as no data is published either on the total value of food aid by commodity or on the share used for different purposes. The estimates are based on data in WFP [1986], FAO [1986] and OECD [1986], supplemented by estimated unit-values of different commodities.
11. For a discussion of this issue see Maxwell (ed.) [1988]; Maxwell and Fernando [1989].
12. In principle, this list could be extended to include GDP as a whole, especially if general equilibrium relationships play an important part in the analysis. It is excluded here in the interests of simplification.
13. This problem of 'adding-up' impacts which occur across levels of analysis and some of which may be qualitative is familiar in other fields. See, for example, Lipton [1985] on the Green Revolution and Maxwell and Fernando [1989] on the cash crops debate.
14. For a review of food security measures, see World Bank [1986a]; Gittinger (ed.) [1988].
15. Maxwell [1986b: 10–11]. These differences could not be explained by differences in definitions or accounting practices.
16. Imported cheese and wine were included, however! For a fuller discussion, see Maxwell [1986b].
17. On rapid rural appraisal, see Chambers [1983]. On participative monitoring and evaluation, see Feuerstein [1983].
18. Data on Senegal from Maxwell [1986b]; on Ethiopia from Maxwell [1986c]; and on Sudan from IDS [1988].
19. Sources for all data quoted are to be found in Maxwell [1986c].
20. Data on producer prices are collected by the Relief and Rehabilitation Commission in 118 markets; a selection of the results is published in early warning reports, for example, Ethiopia [1986]. The data for Woldia are unpublished.
21. This in itself is a major disincentive issue, corresponding to Scenario 2 in Figure 1. See Maxwell [1986c: 17ff].
22. The sources for these figures are to be found in Maxwell [1986b].
23. The failure of the price support operation for millet is an important issue not discussed here but analysed by Maxwell [1986b: 12–15].
24. These and most other figures in this section are taken from IDS [1988].
25. This would require that there be no restrictions on cereal imports which is, admittedly, a brave assumption.

26. The subsidy would have been greater if wheat imports had been valued at the shadow exchange rate, since the Sudanese Pound was overvalued. IDS [1988] used a shadow rate of US$1=LS 6.

REFERENCES

Abbott, P. and F. McCarthy, 1982, 'Welfare Effects of Tied Food Aid', *Journal of Development Economics*, Vol.11, No.1, Aug., pp.63–79
Ahluwalia, I.J., 1979, 'An Analysis of Price and Output Behaviour in the Indian Economy: 1951–1973', *Journal of Development Economics*, Vol.6, No. 3.
Baribeau, R.D. and C.D. Gerrard, 1984, *A Review of Food Aid Literature*, CIDA, Ottawa.
Barlow, F.D. and S.A. Libbins, 1969, *Food Aid and Agricultural Development*, ERS Foreign Agricultural Economic Report No.51, USDA, Washington, DC, June.
Bezuneh, M. and B.J. Deaton, 1981, *Food Aid Disincentives and Economic Development: Some Reconsiderations in Light of the Tunisian Experience*, Virginia Polytechnic Institute and State University (Blacksburg, Virginia), Department of Agricultural Economics Research Report.
Bezuneh, M., B.J. Deaton and G.W. Norton, 1988, 'Food Aid Impacts in Rural Kenya', *American Journal of Agricultural Economics*, Vol.70, No.1, Feb.
Blandford, D. and J. Plocki, 1977, *Evaluating the Disincentive Effect of PL480 Food Aid: The Indian Case Reconsidered*, Department of Agricultural Economics, Cornell University, Ithaca, New York, July.
Buchanan-Smith, M., 1988, 'Grain Marketing in Darfur: Preliminary Report', Masdar UK for Overseas Development Administration and Darfur Regional Government Technical Cooperation Project for Regional Planning, El Fasher, March (mimeo).
Canadian International Development Agency (CIDA), 1984, *Review of CIDA's Counterpart Fund Policy and Practice* (Executive Summary and approved policy decision), Ottawa.
Canadian International Development Agency, 1985, *Corporate Evaluation of Food Aid, Phase I: Integrated Report (Final draft)*, Evaluation Division, Policy Branch, Ottawa.
Cathie, J., 1989, 'Modelling the Role of Food Imports, Food Aid and Food Security in Africa: A Case Study of Botswana' in this volume.
Chambers, R., 1983, *Rural Development: Putting the Last First*, London: Longman.
Clay, E.J. and H.W. Singer, 1985, *Food Aid and Development: Issues and Evidence*, WFP Occasional Papers, No.3, World Food Programme, Rome.
Dandekar, V.M., 1965, *The Demand for Food and Conditions Governing Food Aid during Development*, WFP Studies, No.1, World Food Programme, Rome.
Dudley, L. and R.J. Sandilands, 1975, 'The Side-effects of Foreign Aid: The Case of PL480 Wheat in Colombia', *Economic Development and Cultural Change*, Vol.23, No.2, Jan.
Ethiopia, 1986, *Early Warning System Report: 1985 Meher (main) Crop Seaason Synoptic and 1986 Food Supply Prospect Final Report*, Early Warning and Planning Services, Relief and Rehabilitation Commission, Addis Ababa, Jan.
Food and Agriculture Organisation (FAO), 1961, *Development Through Food: A Strategy for Surplus Utilisation*, Rome.
FAO, 1982, *The Impact of WFP Food Aid in Ethiopia*, Rome, Dec.
FAO, 1984, Committee on Agriculture, *Agricultural Price Policies*, Rome.
FAO, 1986, *Food Aid in Figures* No.4, Rome.
Feuerstein, M.T., 1983, 'Participatory Evaluation: By, With and For the People', in *Rural Life*, Vol.28, No.4.
Fitzpatrick, J, and A. Storey, 1989, 'Food Aid and Agricultural Disincentives', *Food Policy*, Vol. 14, No. 3, Aug.

Gittinger, J.P., et al., 1987, *Food Policy: Integrating Supply, Distribution and Consumption* (EDI Series in Economic Development), Baltimore, MD: Johns Hopkins.

Hall, L., 1980, 'Evaluating the Effects of PL480 Wheat Imports on Brazil's Grain Sector', *American Journal of Agricultural Economics*, Vol.62, No.1, Feb.

Institute of Development Studies (IDS), 1988, 'Food Security Study, Phase 1', report to Government of Sudan, University of Sussex, IDS Feb. (mimeo).

Isenman, P.J. and H. Singer, 1977, 'Food Aid: Disincentive Effects and their Policy Implications', *Economic Development and Cultural Change*, Vol.25, No.2, Jan.

Jackson, T., 1982, *Against the Grain*, Oxford: OXFAM.

Jost S. and J.-J. Gabes, 1988, 'Food Aid to the Sahel: The Situation in 1987/88', Club du Sahel/OECD, Paris, Nov.

Lipton, M, 1985, 'Modern Varieties, International Agricultural Research and the Poor', Consultative Group on International Agricultural Research, Study Paper No.2, Washington, DC: World Bank.

Lipton, M., 1987, 'The Limits of Price Policy for Agriculture: Which Way for the World Bank?' *Development Policy Review* Vol. 5, No.2, June.

Mann, J.S., 1967, 'The Impact of PL480 Imports on Prices and Domestic Supply of Cereals in India', *Journal of Farm Economics*, Vol.49, No.1, Feb.

Maxwell, Simon (ed.), 1983, *An Evaluation of the EEC Food Aid Programme*, IDS Commissioned Study, University of Sussex, No.4.

Maxwell, Simon, 1986a, 'Food Aid: Agricultural Disincentives and Commercial Market Displacement, *Discussion Paper 224*, IDS, University of Sussex, Dec.

Maxwell Simon, 1986b, 'Food Aid to Senegal: Disincentive Effects and Commercial Displacement', *Discussion Paper 225*, IDS, University of Sussex, Dec.

Maxwell, Simon, 1986c, 'Food Aid to Ethiopia: Disincentive Effects and Commercial Displacement', *Discussion Paper 226*, IDS, University of Sussex, Dec.

Maxwell, Simon, (ed.), 1988, 'Cash Crops in Developing Countries', *IDS Bulletin* Vol.19 No.2 University of Sussex.

Maxwell, Simon, (ed.), 1991, *Food Security in Sudan*, IT Publications, forthcoming.

Maxwell, Simon and A. Fernando, 1989, 'Cash Crops in Developing Countries: The Issues, the Facts, the Policies', *World Development* Vol.17, Nov.

Maxwell, Simon, and H.W. Singer, 1979, 'Food Aid to Developing Countries: A Survey', *World Development*, Vol.7, pp.225–47.

Mellor, J.W., 1976, *The New Economics of Growth: A Strategy for India and the Developing World*, Ithaca, NY: Cornell University Press.

Mellor, J.W., 1980, 'Food Aid and Nutrition', *American Journal of Agricultural Economics*, Vol.62, No.5, Dec.

Merrill, A.K., 1977, 'PL480 and Economic Development in Recipient Countries', Washington: National Security and International Affairs Division, Congressional Budget Office (mimeo).

Nelson, G.O., 1983, 'Food Aid and Agricultural Production in Bangladesh', *IDS Bulletin*, Vol.14, No.2, April, University of Sussex.

Organisation for Economic Cooperation and Development (OECD), 1986, *Development Cooperation Review*, Paris.

Raikes, P., 1988, *Modernising Hunger: Famine, Food Surplus and Farm Policy in the EEC and Africa*, London: James Currey.

Rogers, K.D., U.K. Srivastava and E.O. Heady, 1972, 'Modified Price, Production and Income Impacts of Food Aid under Market Differentiated Distribution', *American Journal of Agricultural Economics*, Vol.54, No.2, May.

Schultz, T.W., 1960, 'Value of US Farm Surpluses to Under-developed Countries', *Journal of Farm Economics*, Vol.42, No.5, Dec.

Sénégal, 1984, *Conseil Interministeriel sur la Nouvelle Politique Agricole* (séance du 26 mars 1984), Ministère du Développement Rural, Dakar.

Shuttleworth, G, 1988, 'Grain Marketing Interventions by the State: What to do and why to do it', paper presented to the Workshop on Food Security in the Sudan, IDS,

Sussex, 3–5 Oct. 1988, in Maxwell (ed.), 1991, *Food Security in Sudan*.
Singer, H.W., 1987, 'Food Aid', *Development Policy Review*, Vol.5, No.4.
Stevens, C., 1979, *Food aid and the Developing World*, London: Croom Helm.
Stewart, F., 1986, 'Food Aid: Pitfalls and Potential', *Food Policy*, Vol.11, No. 4, Nov.
Sudan, 1988, 'Proceedings of the National Food Security Workshop 4–5 June 1988', Khartoum (mineo).
Thomas, M, *et al.*, 1989, 'Food aid to Sub–Saharan Africa: A Review of the Literature', *Occasional Paper* No. 13, Rome: World Food Programme.
Timmer, 1986, *Getting Prices Right: The Scope and Limits of Agricultural Price Policy*, Ithaca, NY: Cornell University Press.
US Agency for International Development (USAID), 1983, 'A Comparative Analysis of Five PL480 Title I Impact Evaluation Studies', *AID Programme Evaluation Discussion Paper*, No.19, Washington, DC.
USAID, 1985, *Background Paper and Guide to Addressing Bellmon Amendment Concerns on Potential Food Aid Disincentives and Storage*, Bureau for Food for Peace and Private Voluntary Assistance, Washington. DC.
World Food Council (WFC), 1985a, *Effectiveness of Aid in Support of Food Strategies*, Rome.
WFC, 1985b, *Progress in Implementation of Food Plans and Strategies in Africa*, Rome.
World Food Programme (WFP), 1986, *Review of Food Aid Policies and Programmes*, WFP/CFA: 21/5, WFP, Rome, March.
WFP, 1987, *Agricultural Disincentives*, Rome, Aug. (mimeo).
World Bank, 1986a, *Poverty and Hunger: Issues and Options for Food Security in Developing Countries*, Washington, DC.
World Bank, 1986b, *World Development Report*, Washington, DC.

4

Modelling the Role of Food Imports, Food Aid and Food Security in Africa: The Case of Botswana

JOHN CATHIE

METHODOLOGICAL AND DATA PROBLEMS

The effects of food aid on a recipient economy cannot properly be ascertained without a consistent data base; this is true for micro, macro, sectoral, distributional and nutritional assessments. If food aid is a valuable development resource, it is essential that its contribution either to food security or to general economic development can be analysed and understood.

Food aid policy has been built upon a number of concepts such as additionality, usual marketing requirements (UMRs), fair-price policy and the like, which invariably begs the question what contribution does food aid make. Food aid policy can be seen to embody a different kind of 'theory' (or policy) to that which is generally available from the economic tool box. These food aid theories, or policies, must be tested to demonstrate their superiority over general theories of development and the growth process, in which food aid is a major resource.

Food aid policies, and the implicit theories which support them, are almost pre-paradigmatic (to use a piece of Kuhnian jargon) in their conception, since concepts like additionality, while suitably vague and plausible, are difficult to test in reality. Food aid is considered to comply with the concept of additionality if it is over and above the 'normal commercial demand' for imports by the recipient.

Note of acknowledgement

The author would like to thank the Director of the Agricultural Economics Unit, University of Cambridge, I.M. Sturgess for his helpful and incisive comments on this chapter.

Additionality is primarily concerned with foreign demand rather than home demand in the recipient economy. In practice it is difficult to determine, particularly when concessional sales of food aid are involved, what is a legitimate commercial deal and what is an aid deal. Donors often mix both commercial and aid deals as a package. Normal commercial demand is also a vague notion (see Cathie [*1982*] for a fuller discussion of the food aid and agricultural trade problems associated with food aid policy).

Since the concept of additionality originated as a rule-of-thumb to placate developed country agricultural trade and political interests and dispel their fear of losing markets to competitors, it is not amenable beyond its status as a rule of thumb. Additionality might be considered a convention or guideline, since the decision on what is additional consumption is made on a pragmatic political basis. Additionality as a hypothesis is difficult to test, and operationally it is equally difficult to demonstrate that food aid in a recipient country is in fact additional to demand from home and foreign sources. Additionality cannot be a central concept where food aid is to contribute to the economic growth and development of a recipient. If food aid is to contribute to economic growth, as a development resource, it must do so by contributing through the investment function and not just by increasing consumption. Feeding the hungry is a desirable humanitarian aim but food aid of itself will not necessarily contribute to or sustain the growth process.

It is well known that the availability and quality of data for development planning in many Third World countries is at best uneven and at worst non-existent (see, for example, Stolper [*1966*]). Between unevenness and non-existence, inaccuracy is a major problem facing development planners and those wishing to assess the impact of food aid and food security measures. The lack of data can limit the scope for development planning, and methodology is often determined by the scarcity and inaccuracy of the data base rather than by clearly specified economic policy objectives. In spite of food aid having been available as a continuous development resource for some 35 years (that is, since 1954), there is still no universally accepted method of assessing its role, potential adverse effects and likely contribution. Many countries have received food aid over many years, yet studies of its role in their economic development are few and far between (see Cathie [*1982 and 1989: Ch.2*] for a further discussion of this).

Both multilateral and bilateral food aid agencies pay lip service to commodity aid as a form of investment for development purposes; in practice this form of aid is rarely consistently recorded in the development budget of the recipient. Evaluation of its impact on food policy has tended, more often than not, to focus on an audit of commodity use rather than a more appropriate economic evaluation. This is certainly true of the operations of the World Food Programme (WFP) and the European Community (EC) and is also true for much of Public Law 480 (PL480) food aid. The weakness of this approach was well known in the 1960s, and progress in evaluation technique since then has been very slow indeed, if not non-existent (see Smethurst [*1968–69*] for the notion of evaluation as an audit). The beneficial or negative effects of food aid on many projects and programmes are not known, and the implicit assumption is, if it is not doing the country any good 'it is not likely to do it any harm'.

Recipient country estimates of food aid volumes received often differ from the estimates given by donor agencies. The discrepancies are often so great that they cannot be fully explained by a difference of planning period, between national planning agency and international donor agency recording. For example, the Food and Agriculture Organisation's *Food Aid in Figures* did not record accurately the food aid received in Botswana for the years 1965–85. Similarly FAO's production and trade year books contained inaccurate data for food output and agricultural trade. The reasons for this are not difficult to determine. The agencies in the field are often represented by one individual whose function, rightly, is to see that the food aid is accounted for. The problems of administration of themselves are sufficient to occupy such field officers. International food donors, who compete with each other, often in a particular country, are unhelpful to each other and competition between them can amount to rivalry. This is not always healthy because in some cases they have conflicting objectives. Helping the hungry is one thing, how food aid can do this is quite another. Recourse to head office for policy guidance can also cause problems in a recipient country.

The solution to the lack of a data base for evaluation does not lie in the headquarters of the agencies in Rome, Washington or Brussels gathering data for unspecified policy purposes. That approach has been tried for agricultural trade policy since Hot Springs, and it is doubtful whether it has really contributed to the process of data evaluation for policy formulation. Such aggregate and centralised approaches hardly justify their cost. Moreover they may mask the need for national governments in the developing world to take responsibility for building their own data bases for the purpose of formulating policy relevant to their particular needs. It is significant that the WFP and the EC see the problems of food aid policy management being solved by their assuming responsibility for that aspect of development planning in recipient countries. This approach raises issues of national sovereignty.

The data base on which food aid policy can be built is particularly poor since it has not been the practice to integrate food aid resources into the planning process. Like many of the problems associated with food aid policy, this one has a long history and was recognised as a weakness in the early years. In his classic 1960 article Schultz asked what monetary value should be given to food aid. Should it be valued in terms of the cost borne by the donor or of its contribution to the recipient. The question indicates that the value of food does differ when a comparison is made between donor and recipient costs, by as much as 30 per cent in Schultz's estimation. These questions become particularly pertinent when large volumes of food aid are donated to a particular country or region.

If food aid is to be fully integrated into the budgetary and development planning process of the recipient, it is essential for its volume to be translated into monetary terms, that is to say prices. Recipient governments have good reason to measure the value of food aid in terms of the opportunity costs of substitutable commercial imports of similar foodstuffs. The development and planning budget should show estimates of the savings in fully convertible foreign exchange that the commodity aid provides. Although the foreign exchange savings attributable to food aid do not automatically equal investment in the recipient economy since some (if not all) the food is destined for immediate consumption. Attributing the investment component of the food aid will necessarily involve some arbitrary decisions: for example whether it should go through the recipient's capital account (or Development Fund) or through the current balance of payments account. The existence of this option suggests that the additionality principle may be violated. At least it suggests that the food aid should be valued at the lowest cost of the commodity on world markets. It appears that since the national recipient agencies and international donor agencies do not see the need to integrate food aid into the recipient's national planning and budgeting system this is rarely done.

In Botswana no single national agency records food aid data: neither the Ministry of Finance, the Central Bank, Customs and Excise, nor the Ministry of Agriculture; and the local food aid donors are only aware of what their own agency supplies. Recently the Food Resources Department of the Ministry of Finance and Development has begun to co-ordinate food aid, but it remains to be seen whether it will supply data for the budgetary process. The institutional framework in Botswana indicates a lack of concern for food aid linkages in the economy. Many developing countries regard food aid as a 'residual resource' and this may explain its ghost-like existence as far as the budgetary and planning process is concerned. The data on food aid can be found, provided detective work is undertaken by the analyst.

Evaluating the impact of food aid policy can be problematical. It is hampered by inadequate and inappropriate data, that is to say highly aggregated data unsuitable for planning purposes in the recipient economy. In addition, the foreign exchange value of the aid in question has to be estimated, preferably at world prices, and that value integrated into the budgetary process. How to allocate food aid between consumption and investment purposes can also be a problem. When food aid is not recorded in the national budget, it is also not recorded in the balance of payments statistics. During the 1950s and 1960s a number of problems arose when food aid was sold in a recipient country and the resulting counterpart funds held in the recipient's central bank in an account attributable to the donor agency (see Cathie [1989] for a discussion of the counterpart issues associated with the PL480 programme). The foreign exchange value of the donated food should be recorded in the balance of payments account and the International Monetary Fund (IMF) should ensure that these values are included in its financial analysis and accounts of individual countries. Having established these basic requirements the analysis of the economy-wide effects of food aid in a particular country can begin.

As with all forms of policy analysis, an explicit (or implicit) model has to be adopted as a framework. In the case of food aid, as a development resource, there are a number of possible frameworks. At the microeconomic or project level each food aid project within a country can be analysed in the light of the purpose or aim of the national government or international donor agency. For example, cost-benefit analysis can be applied to projects where the purpose is to contribute, through investment, to a particular sector of the recipient economy. As with project planning there are, of course, a number of estimation problems such as shadow wage rates (particularly with food-for-work projects), the appropriate interest or discount rate, and the appropriate foreign exchange rate. These can be dealt with from the corpus of knowledge that has emerged in cost-benefit literature over the last 20 years (see Cathie [1982]).

Nutrition programmes (food aid for school feeding programmes and for pregnant mothers and their children) need not be assessed as an investment (since the time before the benefits of such aid are seen is considerably longer than in a normal investment project). Applying cost-benefit analysis to nutritional planning, presents a similar time horizon problem to estimating forestry projects – not an easy undertaking. Nutritional policies are more amenable to cost-effectiveness techniques, particularly in countries where the nutritional problem is greater than the resources available for its alleviation. Where either cost-benefit or cost-effective techniques are adopted, the projects should be part of the normal national budgeting process and included in the national budget.

Since cost-benefit techniques are partial they are not really helpful to an understanding of price-dependent, or price disincentive effects. The literature on this vexed question is not particularly conclusive, although, that such an effect can be highly undesirable to a recipient, is almost universally accepted by most food aid specialists, in theory at least if not in practice. The issues surrounding disincentive effects are unlikely to be resolved using partial economic analysis frameworks or models. While disincentive effects are important, a disproportionate amount of time and effort has been focused upon them, trying to prove or disprove their benefits or drawbacks. It is conceivable that a particular government may quite legitimately wish to pursue a price policy in its agricultural sector, using food aid supplies to cause 'displacement effects', that is, to lower prices, because the economy may benefit overall from such a policy (see Cathie [*1989*] for a detailed analysis of this view, using South Korea as a case study).

The type of model an analyst adopts will depend upon the objectives of the study. Where food aid is a small share of national imports, or a small share of GDP, there is little point in an over-elaborate analysis or one that tries to encompass all the sectors of an economy in which food aid is not concerned. A possible basis for deciding upon which model would be appropriate may be the share of food aid in imports and in GDP (see Cathie and Dick [*1987*] and Cathie [*1989*] for a detailed analysis of the role of food aid in Botswana and South Korea, respectively). If food aid constitutes between two and four per cent of GDP, or more than ten per cent of imports, or amounts in per capita income terms to more than $50 a year there is a *prima facie* case for an economy-wide analysis of its role in the development process. These measurements can of course be seen as an estimate, or considered as a first approximation or indicator of economies in which food aid is possibly adding to investment as well as to consumption and is therefore more likely to be contributing to economic growth. They are not a rule-of-thumb. There are a considerable number of food aid recipients (at least 20–30 countries) that fall into these categories. Model selection is obviously influenced by the policy predilections of the analyst, where a particular view is being considered or indeed promoted. All models have defects and are designed for particular purposes and no model has a claim to superiority in all respects over others.

The arguments over which agricultural and food policy strategies are appropriate for the optimum development of sub-Saharan African economies tend to focus upon which of two distinctive sets of policy prescriptions and principles will achieve the best possible growth and welfare outcomes. These prescriptions or principles emphasise either orthodox fiscal monetary policies, together with liberal trade practices, or some degree of state intervention in markets combined with international arrangements to achieve both economic growth and development objectives. The multilateral and bilateral institutions, such as the International Bank for Reconstruction and Development (IBRD) and the IMF, represent the more orthodox, free market policy, end of the spectrum, whereas the Food and Agriculture Organisation (FAO) and the World Food Programme, along with the European Community, the United Nations Conference on Trade and Development (UNCTAD) and the United Nations Development Programme (UNDP), tend to favour policies of direct intervention in national and international markets. These institutions argue for a different set of policy options and priorities for nearly every aspect of economic and development policy. In relation to agricultural and development policy for sub-Saharan Africa, they argue over a wide range of alternative policies. In some cases the difference is a matter of emphasis and practicality as well as resources, and in others it is a matter of principle. Viewing the alternative prescriptions as a tendency towards a set of policies is perhaps more productive than considering them as the outcome of some latter-day ideological morality play (see IBRD [*1988*] and WFP/ADB [*1987*]).

FOOD SECURITY AND NATIONAL FOOD POLICIES

Some of the alternative policy proposals, as they apply to the agricultural and food sectors in sub-Saharan Africa, emerge in the debate over food security and national food policies. The concept of food security, or its obverse, food insecurity, like many generalised notions in economics, embodies a different number of quite separate ideas and policy objectives. At the most general level food security policies are seen as providing 'access by all people at all times to enough food for an active, healthy life' [*Reutlinger, 1985*]. The definition of food security at this level of generality is not helpful in deciding what policies would be appropriate to achieve it. Further examination of the concept indicates that food insecurity can be either transitory or chronic, the latter defined as the persistent existence of malnutrition and the associated lack of development and growth in low income developing economies or in regions of these economies.

Policy recommendations for chronic food insecurity include minimum daily food intake, nutritional policy measures, and often more general, non-economic, measures to offset or improve mal- and under-nutrition. The permanent alleviation of chronic food insecurity is most likely to be achieved by economic growth and development itself and the policies recommended tend to be of a longer-term nature. Transitory (or short-run) food insecurity arises from a temporary decline in a household's, region's, or nation's access to food, arising from a decline in domestic production or an increase in world food prices. Transitory food security policy aims primarily to stabilise trend consumption or fluctuations around the trend, and to do this without affecting the rate of economic growth or the balance of payments. National food insecurity can arise from a number of conditions, including inappropriate food, agricultural and economic policies, which cause instability in overall food production, and fluctuations in international food prices. Individual household food security can be affected by income instability, drought, and supply problems. And within the nation state food insecurity can arise from regional instability.

The IMF and IBRD, through structural adjustment programmes, tend to emphasise policies which deal with transitory food insecurity (through the IMF food import facility to ease balance of payments, and structural adjustment loans to improve, in the medium term, the economic capacity of the agricultural sector or the overall economy). FAO on the other hand recommends policies that aim to deal with both chronic and transitory food insecurity. It also favours global information and early warning systems, regular food aid supplies for basic needs, and regional food stock policies (in addition to national stock policies), as a means of dealing with food problems.

Botswana provides a test case for an analysis of the various internationally recommended policies. Its food problem arises from a number of causes. Overall agricultural production is low and in addition supplies are subject to large yearly fluctuations. Malnutrition is endemic, and has been for some twenty years at least. In addition, its dependence upon South Africa for domestic supplies is regarded as undesirable in the longer term.

THE KIEL BOTSWANA MODEL

To analyse the food security problem in Botswana and to assess the various macro-economic policy options open to the government a Johansen/Orani general equilibrium model was adopted, the Kiel Botswana model.

The model structure pays close attention to micro-economic theory. Its key behavioural equations are derived from the constrained optimisation of neo-classical production and utility functions. These equations emphasise the importance of relative prices and substitution prospects in explaining trade flows and the composition of domestic economic activity. The model is centred around a social accounting matrix flow system (SAM) which includes commodity and factor flows, as well as income distribution between government, regions, and households. It thus allows the tracing of the effects of drought and changes in international commodity prices and capital flows on the patterns of domestic economic activity and household income and consumption, and shows the distributional consequences of alternative policy responses. (For a complete discussion of the model theory and its application to Botswana see Cathie and Dick [1987: 96–157]).

The sources of food insecurity in Botswana were grouped into eleven categories which correspond to national income categories (see Table 1). Each category was analysed in terms of its deviations from trend growth. Food insecurity was defined as a deviation from trend consumption and was considered from the perspective of the economy as a whole. The analysis considered in particular the impact of all the main exogenously determined macro-economic changes. The sources of food insecurity, and the means to lessen that insecurity, were not discussed only from the standpoint of domestic agricultural production and the variability in world agricultural trade prices, but of all the sources of income to the economy. The adjustment in the Botswana economy from the exogenously determined shock, or number of shocks, is necessarily reflected in the economy as a whole (see Figure 1 also for a graphical illustration).

The eleven categories of the sources of food insecurity correspond to national income categories and to IBRD/IMF statistics. Each category was analysed in terms of its deviation from trend growth. That is to say, a time series was used to compute a systematic time path which did not consist solely of a 'usual' trend with a constant growth rate over time but also included medium-term (more than one year) cyclical swings. Economic agents are aware of both elements in the time trend

TABLE 1
TIME SERIES OF EXOGENOUS VARIABLES, DEVIATIONS FROM THE TREND
(IN PERCENTAGE TERMS)

Year	(1) Crop Production	(2) Food Import Price (cif) SAR	(3) Food Import Price (cif) US$	(4) Beef Exports	(5) Beef Export Price (fob)	(6) Mineral Export Price (fob) US$	(7) Gross Fixed Capital Formation	(8) Foreign Labour Earnings in South African Mines	(9) Customs Union Remittances	(10) Total Aid excluding Food Aid	(11) Food Aid
1965	−52.36	−0.19	0.53	16.35	9.16	5.72					
1966	28.14	2.37	14.63	5.74	13.18	3.72	11.38	−12.06			16.43
1967	85.27	7.64	22.58	−3.76	13.66	2.92	−8.17	−3.07			−56.21
1968	−23.14	0.30	−29.44	−13.15	−36.39	−4.74	−13.07	8.73	21.14		79.98
1969	19.41	−2.34	−1.99	−28.25	−1.25	8.71	−37.26	−4.52	−47.99	11.60	40.92
1970	−137.61	−2.39	−12.50	−13.19	−1.34	−26.37	43.21	36.76	23.10	6.47	137.64
1971	38.61	−6.87	−12.33	8.63	−16.10	3.88	3.47	9.14	−10.88	−29.70	−200.47
1972	46.44	−5.78	−17.13	7.35	−17.61	−4.45	29.53	−20.74	−2.58	−10.18	−138.84
1973	−53.74	−1.80	11.28	13.17	12.86	8.03	6.03	1.54	5.77	41.00	36.36
1974	45.15	−3.02	18.85	8.80	17.96	−4.82	13.80	−16.52	23.85	−8.36	−80.59
1975	11.92	−2.52	3.45	0.51	8.39	−13.05	−43.38	−1.47	32.65	−9.80	

Year										
1976	72.44	0.92	−15.04	10.74	−5.58	−5.39	−6.38	−21.86	−17.36	−2.34	52.98
1977	38.98	6.11	12.22	10.93	−2.89	5.50	−33.14	5.12	−27.18	−13.84	21.90
1978	−38.23	6.05	−2.93	−18.15	−17.83	13.78	−6.62	−2.92	−19.09	−10.50	83.99
1979	−182.29	3.13	5.84	15.87	24.98	14.36	8.05	22.09	5.84	3.35	41.51
1980	2.45	7.38	22.45	−18.51	−6.13	9.44	24.69	2.97	18.91	20.94	33.49
1981	84.12	−0.83	1.67	−1.34	12.65	−11.50	19.25	9.10	7.29	−3.78	−2.56
1982	12.54	−0.31	−12.30	−18.29	13.58	9.12	3.29	−6.60	−9.82	21.89	−25.75
1983	18.54	−2.36	−6.45	12.39	6.72	−1.89	4.08	−5.70	−17.54	10.20	−10.58
1984	−16.66	−5.49	−3.41	4.13	−28.01	−12.95	−18.71		2.63	−26.94	−30.19
Standard Deviation	68.32	4.20	14.10	13.50	16.30	10.40	24.20	14.50	21.33	18.41	85.27

Source: Own calculations with the data from Cathie and Dick [1987: Table 3]

The Kiel Botswana Model comprises 1–6 and 12–14, i.e.
(12) Capital Formation in the Mineral Sector (72.04)
(13) Foreign Labour Remittances (19.9)
(14) Net Resource Inflow (41.10)

and only deviations from this systematic path can be regarded as being unexpected and therefore unstable on a year-to-year basis. A systematic time path was calculated from each annual series by estimating with ordinary least squares (OLS) the equations

$$y(t) = a + b_1 T + b_2 T^2 + c_1 y(t-1) + c_2 y(t-2) \qquad [1]$$

$$\ln y(t) = d + e_1 T + e_2 T^2 + f_1 \ln y(t-1) + f_2 \ln y(t-2) \qquad [2]$$

$$(t=1, \ldots, l),$$

choosing the equation with best fit, and picking independent and lagged dependent variables which improved the F-value by at least 0.25. An h-test was used to make sure that the residuals were uncorrelated (white noise). A comparison of each shock, and the relative instability of each category, which is the standard percentage deviation of the actual time series from its systematic time path, was in turn analysed (Table 1).

The single major source of instability facing macro-economic policy in Botswana, seen from column (1), is crop production. Other important sources of macro-economic instability are the beef sector, which has a supply instability of 13.50 (mainly from foot and mouth disease) and a price instability of 16.30, and gross fixed capital formation and customs union remittances with a standard deviation of 24.20 and 21.33 respectively.

The Kiel Botswana model comprises the time series of exogenous variables (1–6 in Table 1) together with three other macro-variable instabilities: capital formation in the mineral sector, foreign labour remittances and net resource inflow (with 72.04, 19.9 and 41.0 deviation in percentage terms respectively) were incorporated into the model. Net resource inflow is a residual factor and the term is used in the Botswana model as a closure factor. It comprises (exports − imports + labour remittances + money investments + customs union remittances) + unrequited transfers − factor payments to the rest of the world + flows of reserves. The instability of this term reflects the instability primarily of capital flows other than direct foreign money investment.

Economic aid from 1969–84, both grants and loans, showed an instability of 18.41 per cent (lower than that of customs union remittances),[1] whereas the supply of food aid in Botswana shows a standard deviation of some 85.28 per cent which is the highest single source of instability to the economy.[2]

Since the country's independence in 1965, total aid has formed a major source of income and investment for the Botswana economy. As a proportion of GDP it was as high as 75.7 per cent in the immediate

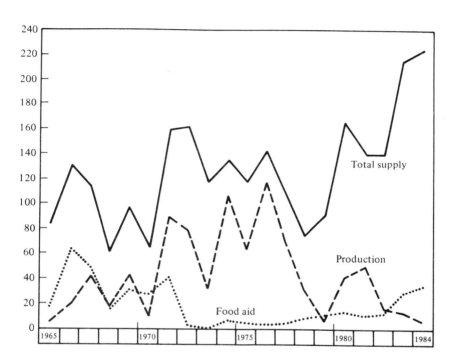

FIGURE 1
CEREAL PRODUCTION AND FOOD AID 1965-84 (1000 mt)

Source: Cathie and Dick [*1987*].

post-independence period. Throughout the period it has sustained development and, together with customs union remittances, largely accounted for total investment, and thus the growth rate, until 1978. Increased investment between 1978 and 1983 (Figure 2) derives from the investment in the diamond sector. Half of the diamond investment is by De Beers and the remainder by Debswana (the Botswana-owned mining company). However, the growth benefits from diamond investments have provided revenue and foreign exchange to government rather than having direct spin offs for the population at large. Similarly the growth benefits of customs union remittances have been to increase government revenue rather than provide immediate income, employment or sustenance to the population at large.

FIGURE 2
RESOURCE INFLOW TO BOTSWANA 1965–1984 (million pula)

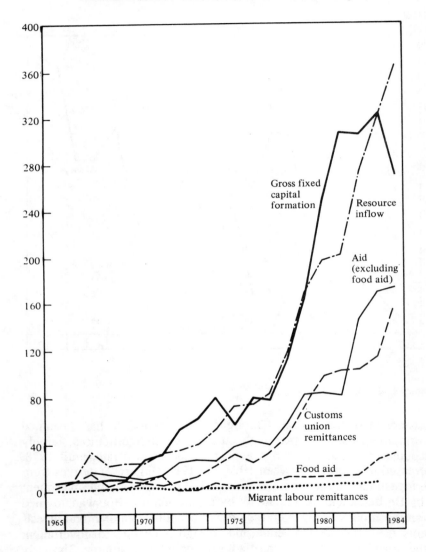

Source: Table 3; Cathie and Dick [*1987*].

Aid, because of its emphasis on building infrastructure, has probably provided more sustained employment over the period than the other resource inflows. The development and growth of the Botswana economy has been fuelled by economic aid which can largely be credited with the 'success', taking the period as a whole. However, aid has not secured self-sustaining development or, for that matter, the spread of benefits to the population as a whole. It has promoted urban development and provided urban infrastructure for the economy.

Both economic aid and customs union remittances have grown in tandem, reflecting in part the political concern of donors in the Southern African region and South African influence on the economic structure of Botswana government revenues. Together, aid and customs union remittances have accounted for one third of GDP since 1966, although the sources of aid have changed over the period [*Stevens, 1981*]. Botswana's economic policy is very much determined by both South Africa's foreign policy objectives and the economic and political aims of the major aid donors.

FOOD AID IN BOTSWANA

Food aid, as indicated by its standard deviation of 85.82 per cent, was extremely uneven if not outright unreliable during the twenty years from 1966–1986. It might have been preferable for the government of Botswana to have had stable food aid supplies or supplies counter-cyclical with production. In the event they were neither. Food aid has not contributed to immediate economic growth because of its overwhelming bias to nutrition programmes, but it has sustained subsistence and ultimately population growth. Per capita food aid has varied considerably over the period, again reflecting donor concerns about periodic drought (see Cathie and Dick [*1987: Table 8 and Table 2*]). From 1966 to 1971, it was a major economic resource, constituting some 18 per cent of GDP. Throughout the period as a whole it has averaged 2–4 per cent of GDP and must be comparable to labour remittances in its effects on sustaining the rural poor. Food aid however perpetuates the subsistence situation whereas labour remittances are a potential source of savings for the poor, and, were it not for cattle purchases, may provide the only source for investment to which they have access. Food aid that is consumed cannot be either an investment, or savings.

Food is given to Botswana by a number of donors, most notably the bilateral agencies of the EC Commission and of the USA as well as the multi-lateral UN food aid agency, the World Food Programme (WFP).

The WFP is a major donor to Botswana and the variation in supplies is related to its difficulties in securing pledges from the agricultural surplus donors. The EC has similar supply problems (see Cathie [*1988*] for a discussion). Both organisations are highly bureaucratic and, as has been previously discussed, decision making is notoriously protracted and subject to administrative assessment, rather than economic efficiency, criteria.

Variation in the supply of food aid to Botswana is partly due to the availability of surplus agricultural produce from donors and partly to drought conditions in the country itself. A regression analysis shows that Botswana has been successful in attracting food aid after the drought period,[3] but that relationship cannot be expected to persist in the future.

In the immediate post-independence period food aid was Botswana's major source of food imports constituting as much as 40 per cent or more of total supply (see Table 2, column 8). Table 2 also shows how the value in pula terms of food aid was calculated on the 'opportunity cost to the import bill' basis. Throughout the 1970s food aid contributed a significant four to five per cent of supply and since the late 1970s its contribution to total supply has been rising. This increase in food aid supplies is now due mainly to the building up of the strategic stock in line with EC/FAO/SADCC (Southern African Development Co-ordination Conference) strategic and reserve stock policy objectives. The initiatives for this policy have come from both the EC and FAO.

The main recipients of food aid commodities are nutritional, school feeding, and vulnerable group feeding, programmes [*Stevens, 1979*]. The programmes are assessed on nutritional rather than economic criteria and they constitute a major source of sustenance for the growing population. Thus the continuous supply of food aid has developed a dependency syndrome in Botswana, with periodic drought reinforcing the continuation of supplies.

A study of the role of food aid in Botswana in 1974 indicated that, as donors had observed the principles of the surplus disposal guidelines followed by the WFP, there was little evidence of food aid causing price displacement effects and, although there were minor expected leakages (arbitrage), overall food aid was having a beneficial nutritional effect on the population [*WFP/Government of Botswana, 1974*].

In the 1970s the government of Botswana, like many other LDCs under WFP guidance, instituted food-for-work programmes in an attempt to relate food and wages to economic development, productivity and growth. In the event these programmes did not prove to be particularly effective as a means to sustained development. The government now considers payment-in-kind (food for wages), as opposed to wage payment for work, to be undignified and no longer practises this in its National Food Strategy (NFS) [*Ministry of Finance and Development Policy, 1985*]. This view is completely in accord with the International Labour Organisation (ILO), Convention 95, which regards payment-in-kind as essentially undesirable and stipulates that no more than 50 per cent of wages should be in kind.

Food aid to Botswana in the 1980s has changed the emphasis of its purpose. While nutritional programmes are to continue, additional food aid surpluses are being stockpiled for both strategic and anticipated drought supply problems. Cereals in storage will amount to some five to six months' commercial import demand with two months of this being earmarked for drought/nutritional purposes.

In practice, the differentiation of this storage capacity into nutritional and strategic purposes will be extremely difficult. Arable agriculture is liable to bear the brunt of adverse secondary effects, with the changed emphasis from a nutritional to a strategic/reserve policy making it increasingly liable to price disincentive effects. Together with the price mechanism bias in the Southern African Customs Union (South Africa, Botswana, Lesotho, Swaziland and Namibia), the food aid storage policy will at best be the substitution of one price disincentive bias for another, yet again undermining the arable agriculture sector and not, as intended, strengthening and developing it in terms of either output or employment.

Food aid policy has sustained and continues to sustain the subsistence poor, particularly through feeding programmes for children. While these objectives are commendable and humane, the government shows no signs of assuming more responsibility for the nutritional well-being of the population, that is to say free from international charity. For some 20 years as much as 50 per cent of the population has depended upon food aid and in 'normal', non-drought, years that dependence has become entrenched with little sign of self-sustained improvement. The growth success in Botswana has not developed the mechanisms, market or otherwise, or the wherewithal, to provide longer-term food security without international food aid, or for that matter without South African subsidies. Both the international charity of food aid and the less than charitable South African subsidies, together with the natural endowment, have perpetuated a poverty trap which does not allow for a prospect of self-sustained access to the necessities of life. Botswana's economic success, when considered against the perpetuated indignities of poverty, charity

TABLE 2
CEREAL PRODUCTION, IMPORTS, AND ESTIMATED VALUE OF FOOD AID
1965–86

Year	Domestic cereal production			Cereal import volume (a)	Food aid volume	Total cereal supply	Food aid volume/ cereal import volume	Food aid volume/ total cereal supply	Cereal import volume/ total cereal supply	Estimated value of food aid (b)	'True' food import bill (c)
	Sorghum	Maize	Total								
			(1)+(2)			(3)+(4)+(5)	(5).100/(4)	(5).100/(6)	(4).100/(6)		
			1000 mt				per cent			Mill. P	
	(1)	(2)	(3)	(4)	(5)	(6)	(7)	(8)	(9)	(10)	(11)
1965	3.2	1.9	5.1	61.0	17.7	83.8	29.02	21.12	72.79	1.46	6.50
1966	18.1	1.4	19.5	48.4	62.7	130.6	129.55	48.01	37.06	6.53	11.57
1967	36.3	5.4	41.7	23.3	49.4	114.4	212.02	43.18	20.37	15.18	22.34
1968	10.4	7.4	17.8	26.7	16.7	61.2	62.55	27.29	43.63	4.12	10.70
1969	29.8	12.8	42.6	23.7	30.6	96.9	129.11	31.58	24.46	8.63	15.31
1970	7.8	2.1	9.9	27.4	26.9	64.2	98.18	41.90	42.68	7.42	14.98
1971	73.3	16.6	89.9	28.7	40.3	158.9	140.42	25.36	18.06	13.76	23.56

BOTSWANA: FOOD IMPORTS AND FOOD SECURITY

Year											
1972	68.3	10.3	78.6	80.0	2.7	161.3	3.38	1.67	49.60	0.49	15.05
1973	10.3	22.3	32.6	85.0	0.7	118.3	0.82	0.59	71.85	0.15	18.38
1974	72.4	33.9	106.3	22.0	6.5	134.8	29.55	4.82	16.32	6.26	27.44
1975	33.8	28.7	62.5	50.0	5.4	117.9	10.80	4.58	42.41	3.13	32.12
1976	55.5	62.6	118.1	20.9	3.4	142.4	16.27	2.39	14.68	5.78	41.32
1977	33.0	35.4	68.4	34.0	4.7	107.1	13.82	4.39	31.75	6.36	52.33
1978	15.5	14.0	29.5	36.8	8.4	74.7	22.83	11.24	49.26	12.18	65.56
1979	4.3	2.2	6.5	72.2	11.1	89.8	15.37	12.36	80.40	11.05	82.92
1980	29.1	11.6	40.7	109.3	14.9	164.9	13.63	9.04	66.28	11.51	95.90
1981	28.3	21.4	49.7	79.4	10.9	140.0	13.73	7.79	56.71	12.98	107.53
1982	3.9	12.4	16.3	111.0	12.6	139.9	11.35	9.01	79.34	13.15	129.02
1983	5.2	8.5	13.7	173.6	28.3	215.6	16.30	13.13	80.52	25.98	185.34
1984	5.7	0.5	6.2	183.0	34.5	223.7	18.85	15.42	81.81	31.47	198.42
1985	—	—	—	—	66.0	—	—	—	—	—	—
1986	—	—	—	—	25.7	—	—	—	—	—	—
Average	—	—	—	—	—	—	49.38	16.74	49.0	—	—

Notes: (a) Includes food aid; (b) The value of food aid is estimated as the 'opportunity costs to the import bill' by: value of food imports food aid volume/cereal import volume; (c) The 'true' value of the food import bill is calculated as the sum of the value of food imports and the estimated value of food aid.

Source: Eele and Funk [1985]; FAO, a; World Bank [1981; 1986a]; Hay et al. [1986] WFP/Government of Botswana [1974]; Cathie and Dick [1987; Tables 8 and 9].

and being beholden to South Africa, appears a diminished achievement.

Both food aid and the customs union, as well as aid in general, have given expedient resources to the government but have not provided the means for self-sustained development particularly in arable agriculture.[4] The actual continuation of the South African economic dependence and the proposed FAO alternative are likely to maintain the rural *status quo* in Botswana.

THE MODEL RESULTS

Having established the basic macro-economic instabilities that affect food security in Botswana it is possible to analyse the short run policy options facing the government in response to a drought. There are three other short run policy simulations for 1981/82 analysing international prices, net resource inflows (balance of payments) and an all shock Cassandra simulation.

The results of the drought simulation for the year 1981/82 are given in Table 3. All values, with the exception of the balance of payments which is in million pula, are percentage changes. The overall effect of the drought is to reduce GNP by five per cent (that is, benchmark simulation 4).

Simulations 1–3 consider the government's options when it wishes to maintain formal economic growth (that is to say real domestic absorption is held constant). If it wishes to maintain consumption in poor households in the regions it has a choice of 'nutritional transfers' (essentially food aid to schools and vulnerable groups), supplementing the full cost of the food aid itself. Regional transfers for both consumption and investment purposes go to local authorities, whereas nutritional transfers (feeding programmes) do not add to growth in the short run; at least this is not measurable.

The three policy simulations 1–3 indicate that, for the distribution of food consumption to poor households in the regions of Botswana, a *food price stabilisation policy has the worst results*. Price stabilisation is a preferred policy option of the interventionist multilaterals arising from the promotion of emergency or strategic stock policies. These stocks, it is argued, will provide for the poor during periods of shortage, and therefore have direct distributional objectives (improving consumption) which override indirect distributional objectives (improving producers' incomes) such as appropriate price policies for rural food producers.

Simulations 4–8 cover government policy options if the balance of payments is held constant. Simulation 4 is the benchmark simulation to which other policy options can be compared. While Botswana has a good monetary reserve position compared with many developing countries, there is evidence (net resource inflow instability of 41.0 and the financial crisis of 1981 due to the collapse of the diamond market) that this reserve position is fragile.

Simulations 6 and 7 maintain the food consumption of the poor in the regions, but result in growth being reduced by a further four per cent compared with simulation 4.

Simulation 8, cutting government consumption and stabilising food prices, affects growth in a similar way to doing nothing at all (simulation 4, comparison) but the distribution of food consumption to households in the regions is much the same as doing nothing at all.

These simulation results indicate that the price stabilisation policy option will not achieve the intended purpose. The results from the other simulations (international price, net resource inflow, all shocks) indicate a similar conclusion (see Cathie and Dick [*1987*]).

The government of Botswana in its national food strategy has established food stocks of six to eight months' consumption to meet drought and other contingencies. Using stocks with a price policy is unlikely to improve the food consumption of the poor in the regions. Indeed, it may very well make the situation worse. The donor community may have given Botswana a White Elephant.

CONCLUSIONS

It has been argued that food aid policy modelling involves both methodological and data problems which often limit its scope. This chapter has made a number of suggestions as to how the data base for, and the methodology currently applied to, food aid policy formulation might be improved. The different definitions of food security as a concept, that is to say transitory and chronic, adopted by the multilateral donor agencies, need to be reconciled. This is particularly so if the IBRD and the WFP wish to co-operate and co-ordinate their aid and food aid policies in recipient economies.

The policy options for food security in Botswana, recommended by the multilateral institutions, have been analysed. The intervention proposals at best will have little effect upon the prospects of the poor and at worst may affect them adversely. Since price stabilisation policy will not improve the distribution of food consumption to households it may also discourage production in the post price stabilisation period. The recommendations of the orthodox multilateral institutions will

TABLE 3
SHORT-RUN POLICY OPTIONS IN RESPONSE TO DROUGHT:
MODEL SIMULATION RESULTS FOR THE YEAR 1981–82[a]

Policy Simulations	(1)	(2)	(3)	(4)	(5)	(6)	(7)	(8)
	Real Domestic Absorption Constant					Balance of Payments Constant		
	Nutritional Transfers	Regional Transfers	Food Price Stabilisation	Allowing Food Consumption to Fall	Allowing Real Govt. Investment Cuts	Cutting Govt. Consumption: Increasing Nutritional Transfers	Cutting Govt. Consumption: Increasing Regional Transfers	Cutting Govt. Consumption: Food Price Stabilisation
				Macroeconomic Policy Indicators				
Real domestic absorption	0.00 (Ex)	0.00 (Ex)	0.00 (Ex)	−5.13	−4.70	−9.81	−8.95	−5.46
Real government expenditure	16.95	11.00	0.02	−2.77	−2.86	−3.01	−2.95	−2.95
Real government consumption	0.00 (Ex)	0.00 (Ex)	0.00 (Ex)	−4.96	0.00	−57.32	−47.97	−5.28
Real government investment	0.00 (Ex)	0.00 (Ex)	0.00 (Ex)	0.00 (Ex)	−9.72	0.00 (Ex)	0.00 (Ex)	0.00 (Ex)
Government expenditure on household and regions	140.18	90.97	0.00 (Ex)	0.00 (Ex)	0.00 (Ex)	243.92	200.45	0.00 (Ex)
Ratio of government expenditure to revenue (Government deficit)	18.56	11.46	0.66	0.00 (Ex)	0.00 (Ex)	0.00 (Ex)	0.00 (Ex)	0.00 (Ex)
Balance of payments (absolute change in reserves, Mill. Pula)	−34.59	−33.89	−36.01	0.00 (Ex)	0.00 (Ex)	0.00 (Ex)	0.00 (Ex)	0.00 (Ex)

BOTSWANA: FOOD IMPORTS AND FOOD SECURITY

Distribution of Food Consumption to Poor Households in the Regions

Southern	0.00 (Ex)	0.00 (Ex)	-37.85	-39.17	-38.24	0.00 (Ex)	0.00 (Ex)	-39.21
Central	0.00 (Ex)	0.00 (Ex)	-11.54	-14.67	-13.63	0.00 (Ex)	0.00 (Ex)	-13.62
Maun	0.00 (Ex)	0.00 (Ex)	-36.78	-38.86	-38.22	0.00 (Ex)	0.00 (Ex)	-37.92
Western	0.00 (Ex)	0.00 (Ex)	3.69	0.50	1.03	0.00 (Ex)	0.00 (Ex)	2.11
Urban	0.00 (Ex)	0.00 (Ex)	-2.16	-5.83	-4.75	0.00 (Ex)	0.00 (Ex)	-4.42

Price and Quantities of Food

General price for arable agricultural	3.20	3.20	0.00 (Ex)	2.80	2.82	3.13	3.07	0.00 (Ex)
Foreign and Domestic Food Price Difference	0.00 (Ex)	0.00 (Ex)	-3.10	0.00 (Ex)	0.00 (Ex)	0.00 (Ex)	0.00 (Ex)	-3.06
Arable agricultural commodity import volume (percentage terms)	30.41	30.33	20.80	16.17	29.22	26.92	19.42	

[a] Drought (68.32 per cent shortfall in actual output of arable agriculture). With the exception of the Balance of Payments, which is in million pula 1981/82, all values are percentage changes for the year 1981/82.

Ex (Exogenously given)

Source: Own calculations on the basis of data described in Appendix II [*Cathie and Dick, 1987*].

allow maximum food security with the maximum possible rate of economic growth in Botswana. Food aid can provide temporary assistance in the government's nutritional programme, but it should not become an excuse for the neglect (that is to say maintaining the *status quo*) of the subsistence poor or of arable agriculture. The basic problems facing the government of Botswana cannot be solved by food aid, and it has yet to be shown how this form of aid can contribute to the economic growth of the country. The provision of food aid for reserve stocks will not create economic growth and may, if it encourages inappropriate agricultural price policies, retard it.

NOTES

1. Aid (excluding food aid), million pula:

		Estimated Coefficient	Standard Error
Dependent variable	$\ln y(t)$		
Independent variable	T	0.3294	0.0977
	T^2	−0.0130	0.0038
	$\ln y(t-1)$	0.4915	0.2974
	$\ln y(t-2)$	−0.2576	0.2455
Constant		1.4870	0.5662

$R^2 = 0.9473$ $F = 49.33965$ $DW = 2.0159$

2. Food Aid, million pula:

		Estimated Coefficient	Standard Error
Dependent variable	$\ln y(t)$		
Independent variable	T	−0.3569	0.2043
	T^2	0.0242	0.0107
	$\ln y(t-1)$	0.1406	0.2638
	$\ln y(t-2)$	−2.2705	0.2358
Constant		2.7156	1.0736

$R^2 = 0.5013$ $F = 3.2672$ $DW = 1.9908$

3. In the regression for the periods 1965–1984, food aid volume ($\ln y(t)$) shows as a function of lagged crop production ($\ln x(t-1)$, $\ln x(t-2)$)

		Estimated Coefficient	Standard Error
Dependent variable	$\ln y(t)$		
Independent variable	$\ln x(t-1)$	−0.7688	0.1875
	$\ln x(t-2)$	−0.3114	0.1897
Constant		6.1721	0.7550

$R^2 = 0.6151$ $F = 13.5846$ Degrees of freedom: 17

4. See Hudson [*1981*]. This is a good example of the myopic reasoning that government has adhered to in thinking that Botswana is better off inside than outside the customs union. Hudson's analysis relies on administrative and regulatory procedures which provide the government with revenue. He chooses to ignore the economic costs and benefits of the customs union in Botswana.

REFERENCES

Cathie, J., 1989, *Food Aid and the Industrialisation of South Korea 1945–1975*, Aldershot: Gower.

Cathie, J., 1988, *European Food Aid Policy a Review and Assessment 1968–1985*, Agricultural Economics Unit of the Department of Land Economy, Cambridge, Occasional Paper Series No.38.

Cathie, J., 1982, *The Political Economy of Food Aid*, Aldershot/New York: Gower/St Martins.

Cathie, J. and H. Dick, 1987 'Food Security and Macroeconomic Stabilisation: A Case Study of Botswana', Tubingen: *Kieler Studien* 208; J.C.B. Mohr (Paul Siebeck) and Boulder, CO: Westview Press.

Cathie, J. and R. Herrman, 1988, 'The South African Customs Union: Cereal Price Policy in South Africa and Food Security in Botswana', *Journal of Development Studies*, Vol.24, No.3 April.

Eele, G.J. and P.E. Funk, 1985, *Government of Botswana Updating a Food Accounting Matrix (FAM) for Botswana*, Final Report, Oxford: Food Studies Group, Queen Elizabeth House.

Food and Agriculture Organisation 1984, 1986, *Food Aid in Figures*, Rome.

Food and Agriculture Organisation, *Production Yearbook* (various issues), Rome.

Food and Agriculture Organisation, *Trade Yearbook* (various issues), Rome.

Hay, R., S. Burke, and D. Dako, 1986, *A Socio-Economic Assessment of Drought Relief in Botswana*. Gaborone: UN Development Programme (mimeo).

Herrman, R., and J. Cathie, 1987, *Agricultural Price Policy in the Republic of South Africa: The South African Customs Union and Food Security in Southern Africa*, Kiel University Working Paper 289, Institut für Weltwirtschaft.

Hudson, Derek, 1981, 'Botswana's Membership of the Southern African Customs Union,' in Charles Harvey (ed.), *Papers on the Economy of Botswana*, London: Heinemann, pp.131–58.

International Bank for Reconstruction and Development (IBRD), 1981, *Accelerated Development in Sub-Saharan Africa – An Agenda for Action*, Washington, DC.

International Bank for Reconstruction and Development, 1986, *Financing Adjustment with Growth in Sub-Saharan Africa 1986–91*, Washington, DC.

International Bank for Reconstruction and Development (IBRD), 1986, *World Development Report 1986*, Washington, DC.

International Bank for Reconstruction and Development, 1988, *The Challenge of Hunger in Africa: A Call to Action*, Washington, DC: World Bank.

Johansen, Leif, 1974, *A Multisectoral Study of Economic Growth*, Amsterdam: North Holland.

Ministry of Finance and Development Policy, 1985, *Rural Development Council, National Food Strategy (NFS) Report*, Gaborone: MFDP.

Reutlinger, Shlomo, 1985, 'Food Security and Poverty in LDC's, *Finance and Development*, Vol.22, No.4.

Schultz, T.W., 1960, 'Value of US Farm Surplus to Underdeveloped Countries', *Journal of Farm Economics*, Vol.42, No.5.

Smethurst, R.G., 1968/69, 'Direct Commodity Aid: A Multilateral Experiment', *Journal of Development Studies*, Vol.5, No.3.

Stevens, Christopher, 1979, *Food Aid and the Developing World: Four African Case*

Studies, London: Croom Helm.

Stevens, Michael, 1981, 'Aid Management in Botswana: From One to Many Donors', in Charles Harvey (ed.), *Papers on the Economy of Botswana*, London: Heinemann, pp.159–77.

Stolper, W.F., 1966, *Planning without Facts*, Cambridge, MA: Harvard University Press,

World Food Programme/African Development Bank, 1987, *Food Aid for Development in Sub-Saharan Africa*, Rome.

World Food Programme/Government of Botswana, 1974, 'Report on a Study of Constraints on Agricultural Development in the Republic of Botswana (including an assessment of the role of food aid)'. Rome.

5

Dairy Aid and Development: Current Trends and Long-Term Implications of the Indian Case

MARTIN DOORNBOS, LIANA GERTSCH AND PIET TERHAL

INTRODUCTION

Operation Flood, the large-scale Indian dairy development programme, aims at replicating all over the sub-continent a successful type of dairy cooperative which originated in Kheda district of Gujarat. The programme is based on institutional and technological innovations directed at linking cooperatively organised rural milk producers to the main urban markets for milk and dairy products. The programme has been supported by considerable amounts of dairy food aid from the EC since 1970 and has become the largest single dairy aid programme in the world. It is widely considered to be a test case for the feasibility and suitability of using the dairy surpluses of developed countries to develop the dairy sector in Third World countries.

In the wake of accomplishing its agenda, a host of socio-economic benefits were expected to follow. Hailed as a White Revolution, Operation Flood emerged as a policy which would benefit both rural producers and urban consumers. The agencies responsible for implementing it were the National Dairy Development Board (NDDB), which served as an advisory body, and the Indian Dairy Corporation (IDC), which handled the finances of the scheme, including the receipt of dairy aid. The former was a parastatal agency technically operating in the private sector, while the latter was a public sector agency. In 1987, the IDC was merged with the NDDB.

Note of acknowledgement
The authors wish to express their thanks to Marc Verhagen for his many constructive suggestions on their earlier draft of this chapter.

Operation Flood, partly because of its size and its trend-setting role, must be seen in the context of wider changes in the Indian economy, in particular the changing role of bovines, the increased productivity of agriculture, population pressures and urbanisation, and the concomitant growth of urban purchasing power. In the course of time, the two problems of raising Indian milk production and supplying milk to Indian cities became intimately linked in the view of policy makers, who visualised Operation Flood as a way to solve both problems simultaneously. Economic and political factors led to the perception of Operation Flood (with all its technological and institutional implications) as the main (if not the only) dairy development strategy in India. Concomitantly, alternative perceptions were discarded.

From the donor side, while the compulsion of surplus disposal and the interests of export promotion through aid were certainly the basis of EC dairy aid, other motivation became increasingly important, particularly in the case of Operation Flood. The project came to be seen as an ideal example of how food aid could promote indigenous development. The use of aid was partly facilitated by generous reference to the Indian concept of 'the cooperative Anand pattern', which conjures up images of ultimate rural autonomy and suggests that rural producers make and live by their own decisions rather than those of others. Access to this new space for individual autonomy has been seen as a means of promoting greater social equity in that its benefits are ostensibly scale neutral. The evidence of the replicated model, however, reveals varying degrees of social differentiation, with weaker economic households benefiting less than stronger ones. In other respects also the aid-induced dairy development engineered through Operation Flood appeared to be in reality less successful than projected [*Doornbos et al., 1990*]. At the end of Operation Flood II, the European Commission was hesitant to continue dairy aid for the following phase, Operation Flood III, but was pressed by the Indian Government to renew its commitment. While its final decision was indeed to maintain dairy aid to India, a number of conditions were attached to the new round.

This chapter explores some of the long-term implications inherent in the model of dairy development adopted, and, in doing so, it describes the mechanics of some of the strategies employed. But, this score-sheet of performance is overlaid by an interest in the 'instruments of development' whose role is to catalyse, implement and sustain the process.

In the current case, one vital instrument is the NDDB. The public intent of food aid is to bring about food security, be it by strengthening a particular sector of the market, tailoring policies to the nutritional needs of vulnerable groups, or enabling balance of payment support [*Lipton, 1985: 60*]. Yet the possibility of accomplishing any or all of the above requires an institutional infrastructure that can move the necessary social levers at all levels, from government bureaucracies to the grassroots. The NDDB centralised this capacity in its institutional framework and its application of nation-wide constructs to incorporate milch cattle, producers and processing and marketing structures.

The chapter begins with a description of the institutional model and its mode of replication. Next, the strategies of procurement and production are described (the National Milk Grid and the National Herd). These structures point to distinctive features of the programme. The section on the foreign connection which follows discusses the motivations for giving and receiving aid, the uses of aid and its implications. Issues of access and social benefits which constitute the next section also provide an indirect perspective on aid. More resources do not guarantee changes in the patterns of access. Questions of who benefits, to what degree, and has access to what resources, also arise.

Long-term constraints and opportunities are prominent themes in the subsequent two sections dealing with participation and side effects. The first is a look at the balance of power within dairy cooperatives and whether they can indeed act as agents of social change. The second is concerned with the longer term effects of Operation Flood which are not directly addressed in project documents or evaluation reports. The side effects are indirect consequences of intensified dairying which have an impact on the whole agrarian economy. Lastly there are some observations on the repercussions of aid on the institutions that implement development. In this section it is suggested that aid is responsible in part for certain characteristics of the project-implementing agency (centralisation, standardisation, accelerated pace of development).

THE MODEL: ITS ORIGIN AND REPLICATION

The model chosen by NDDB to organise dairying nation-wide was a successful cooperative structure in Kheda district of Gujarat, officially registered as of 1946. Known as the 'Anand Pattern', it is a three-tiered vertical structure: producers group into dairy societies, a number of which form a district union and all the unions within a state form a state federation. The logic of such an arrangement is to perform all dairy sector functions within the co-operative and eliminate the potentially exploitative role of intermediaries, whose interests lie outside the welfare of the dairy producers. The division of tasks among the three tiers is: (a) village societies organise individual producers into a production base; (b) district unions organise milk collection, processing and product manufacture, which entails transport and storage facilities as well as processing and marketing equipment; (c) state federations market the dairy products and ensure that the dairy supply is balanced geographically. The federation also investigates opportunities for new product lines according to its market information. Integration of the whole structure has recently been extended to a fourth tier, the National Cooperative Dairy Federation of India (NCDF). From its vantage point at the top, the NCDF can coordinate the activities of state federations and manage the dairy grid in such a way that, despite seasonal swings, supply is balanced all over the country throughout the year.

Village producers own the union's plant and assets. In return for supplying milk to the co-operative, producer-members receive a fixed price as well as cattle feed at a reasonable cost, veterinary services either free or for a nominal charge and help in obtaining finance from banks or other agencies. Farmer training and the extension of cross-breeding technology are also part of the input package. At each level representatives are elected to the next higher tier to ensure that policy decisions are influenced by input from the base. To augment the roles of owners and representatives, at union level and above the co-operative hires professional managers to inform and guide policy. As employees of the co-operative, managers are accountable to the elected board in the union or state federation in which they work. In this way managerial expertise is used at the discretion of elected co-operative boards.

The advantage of co-operative organisation as envisaged by the NDDB is that all operations relating to a particular commodity are vertically integrated from producer to consumer. Through this structure inputs and extension services are made available to producers. Moreover, the value added at various stages of processing and manufacture is kept within the co-operative so that profits may end up as increased dividends for members and/or bonuses and/or for reinvestment. Individual producers are expected to benefit from economies of scale at all stages of product preparation despite the single producer's very small base of operations. But rather than representing a universal norm in India, the 'Anand pattern' co-operative model should be seen as an ideal from which, in reality, many deviations are found, particularly in relation to 'ownership', decision-making and representation.

In the course of organising dairy producers and resources to increase production and implement a long-distance milk supply system, NDDB/IDC made plans to create a National Milch Herd and a

INDIA: DAIRY AID AND DEVELOPMENT

National Milk Grid. The former, under the second phase of Operation Flood (OF II), was meant to comprise cows crossbred with high-yielding exotic breeds and upgraded buffaloes in order to increase milk production. Fourteen million crossbred cows and upgraded buffaloes, representing one sixth of India's milch animals and yielding an estimated one fourth of milk production, were to be included. The National Grid's purpose was to even out differences in the supply of milk due to seasonal fluctuations and regional variations in production. The grid would allow consumers throughout India to enjoy a stable supply of milk throughout the year. This last structure entails a highly technical network of tankers, chilling plants, feeder balancing dairies, urban dairies and bulk vending outlets.

An important aspect of the implementation of Operation Flood (which implied the replication of Anand type co-operatives all over India) was the relation between NDDB and the state governments. At a certain point the Anand pattern was extended from a two-tier to a three-tier arrangement which implied the need to integrate the co-operative with existing state dairy departments. Heretofore dairying had been a state matter. Shifting dairy authority from a ministry official to an elected co-operative chairman, plus the implications for the other public dairy personnel, created friction. The restructuring meant a withdrawal of state involvement in the dairy sector.

Once the co-operative is accepted in principle, there is further negotiation between the state and the NDDB over a plan for dairy development written by the state but in conformity with Operation Flood objectives. When the IDC approves the plans it makes resources available. The new state federation (or other temporary implementing agency) can apply for a loan if it has sufficient assets as collateral. If not, the state is expected to guarantee the loan. The state is not allowed to audit the federation through its own institutions but is required to establish an independent auditing board. According to the agreement both producer and consumer pricing policy is at the sole discretion of the co-operative, but in practice prices are established in consultation with state officials and in conformity with overall central pricing policy. One further aspect of the agreement requires the state to transfer public rural dairies to the co-operative sector. As a result, state governments at times feel that they have been politically pressured into conforming to the needs of a centralised programme, rather than arriving at a compromise which acknowledges their autonomy and specific priorities.

PROCUREMENT AND PRODUCTION

Evidence on the rural impact of Operation Flood is not always reliable and is further obscured by: (1) the impossibility of isolating the impact of Operation Flood from other factors operating in the Indian rural economy and, (2) the very large diversity of ecological and socio-economic conditions under which Operation Flood has beem implemented. The reality of the dairy programme is probably much more mixed in terms of success and failure than either project authorities or various commentators have been willing to acknowledge.

Success in meeting a number of targets can be deduced – on the basis of official records – from the number of co-operatives and participating farmers and the volume of milk collected. All three indicators expanded considerably according to the records provided by the project authorities. Compared with the scaled-down targets of the World Bank Appraisal Report, and particularly the Sixth Plan, achievements are not bad (see Table 1). However, underlying these seemingly satisfactory quantitative achievements are notable shortcomings in meeting various others targets, particularly with respect to the input programme (veterinary services, artificial insemination, animal feed, see Table 2) and the democratic functioning of the co-operatives [*Doornbos et al., 1990: Ch.4*].

Spending under Operation Flood appears to have been biased towards hardware for milk processing and marketing, and to have favoured the more 'successful' states in the west of the country (see Tables 3 and 4). In this connection, the apparently special position within the national dairy development strategy granted by the NDDB/IDC first to AMUL (the name of the dairy in Anand) and next to the Gujarat Cooperative Milk Marketing Federation, is noteworthy. Given the level of investment in Gujarat, it is not surprising that it has the greatest number of dairy co-operatives and is responsible for the highest levels of milk procurement [*Doornbos et al., 1990: Ch.4*]. However, within a policy of replicating a standard package, it is difficult to justify the special privileges accorded to AMUL, the model.

The scale of equipment installed by Operation Flood has often led to underutilisation, particularly as milk production has not met optimistic expectations. The original production and procurement targets were highly ambitious and subsequently required scaling down [*Doornbos et al., 1990: Ch.4*]. An NDDB [*1983*] projection suggested that OF II could potentially achieve an overall growth rate of 100 per cent in total milk production in India between 1978 and 1988. This has not

TABLE 1
OPERATION FLOOD II – TARGETS AND ACHIEVEMENTS AS RECORDED IN DIFFERENT PLAN DOCUMENTS

	Original 1978 targets	World Bank Appraisal Report targets (1985)	Sixth Plan targets (1985)	Achiev. up to Sept. 1985	Achievement as % of Original 1978 targets	Achievement as % of World Bank Appraisal Report targets	Achievement as % of Sixth Plan targets
Total no. of village societies	30,000	25,000	29,000	39,486	131	142	136
Societies under A.I.	30,000	25,000	8,000	8,254	27	30	103
Households covered	10,200,000	4–5,000,000	3,480,000	3,995,000	39	89	115
Average milk procurement (mlpd)	18.30	8	5.53	5.60[1]	31	52	101
Urban milk marketing (mlpd)	12.40	9.5	4.30	5.01[1]	40	46	117
Rural dairy processing capacity (mlpd)	14.00	8.0	7.60	9.50	68	119	125
Number of milksheds	115	–	–	146	94	–	–

Source: NDDB, OF Phase III, Sept. 1985, A proposal.
NDDB, OF II, June 1978, A proposal, Table I.
IDC, Quarterly Progress Report, July–Sept. 1985
World Bank Staff Appraisal Mission Report, 1978.
Court of Auditors' Report of the European Communities on Operation Flood II (1988).

Note 1: Average for April 1984 to March 1985.
mlpd = million litres per day
A.I. = artificial insemination

TABLE 2
COVERAGE OF VETERINARY SERVICES, ARTIFICIAL INSEMINATION
AND PROVISIONS FOR CATTLE FEED UNDER OF II (SEPT. 1985)

Dairy Cooperatives existing as of Sept. 1985	Veterinary service		Artificial insemination		Calves born of upgraded females as of Sept. 1985 (1,000)	Balanced feed for cattle				
	Dairy Cooperatives covered as of Sept. 1985	%	Dairy Cooperatives with artificial insemination as of Sept. 1985	%		Plant capacity in tonnes per day (Sept. 1985)	Tonnes produced 1984–85	% of annual plant utilisation[1]	Dairy Cooperatives selling feed as of Sept. 1985	
									number	%
39,486	23,407	59	8,254	21	760,6	3,289	384,884	47	19,591	50

1 Operating 250 days.

Source: Court of Auditors' Report of the European Communities on Operation Flood II (1988).

INDIA: DAIRY AID AND DEVELOPMENT

TABLE 3
DISTRIBUTION OF FUNDS UNDER OF I AND OF II

	Allocation		Disbursement	
	OF I	OF II	OF I	OF II
1. Increase in processing and marketing capacities	55%	42%	64%	61%[1]
2. Increase in milk supply (production, village co-op organisation	36%	41%	26%	24%
3. Supplementary items (national grid, project planning and implementation)	9%	17%	10%	14%

Source; Calculated from Table 4.6 (OF II) in Doornbos *et al.* [*1990*].
Terhal and Doornbos [*1983*] (OF I).
[1] Including expenditure from central equipment pool.

happened. Projections of this kind often overlook the fact that Operation Flood's share in overall Indian milk production (that is, that part which is being procured and processed through the co-operative OF structure,) is about six to seven per cent of the total.

A juxtaposition of official data on Indian milk production and the estimated trends in the demand for milk and milk products would point to a comfortable domestic supply situation [*Doornbos et al., 1990*]. Operation Flood has often been credited with this satisfactory balance, an assessment which *a priori* tends to contrast with the programme's continued dependence on imported dairy products. However, official production statistics have not been very reliable and Operation Flood's contribution the to milk production increases which have taken place tends to be exaggerated (see Baviskar and Terhal [*1990*].

Over the years, a gradual increase of a few per cent per annum is noticed in total Indian milk production. Gradual shifts in the size and composition of the bovine population, that is, the increased number of milch animals as part of the total, particularly in the Green Revolution belt of the country (Punjab, Haryana, Uttar Pradesh), have probably been mainly responsible for this [*Doornbos and Nair, 1990*]. Apparently milk producers are not insensitive to the market incentives arising from the increased urban demand for milk and respond to these

TABLE 4
CUMULATIVE DISBURSEMENT OF FUNDS UNDER OF II
OVER VARIOUS STATES UP TO MID-1986 (MILLIONS Rs)

Goa	13.73
Gujarat	667.85
Madhya Pradesh	388.68
Maharashtra	269.56
Total Western Region	*1,339.82*
Andaman & Nicobar	3.31
Assam	42.28
Bihar	67.30
Nagaland	0.10
Orissa	120.07
Sikkim	8.58
Tripura	3.20
West Bengal	120.17
Total Eastern Region	*365.01*
Andhra Pradesh	380.77
Karnataka	103.85
Kerala	165.29
Pondicherry	9.77
Tamil Nadu	307.43
Total Southern Region	*967.11*
Delhi	81.69
Haryana	93.83
Himachal Pradesh	9.87
Jammu & Kashmir	5.53
Punjab	363.05
Rajasthan	122.98
Uttar Pradesh	142.64
Total Northern Region	*819.59*
Other disbursements	126.12
Total	3,617.65

Source: IDC Progress Reports (various issues)

Note: The category 'other disbursements' includes disbursements for Mizoram, for the Institute of Rural Management (Anand) and – most importantly – NDDB itself (Rs 99.2 million).

by effecting gradual changes in the composition of their bovine holdings, that is by keeping slightly more (indigenous) milch animals of a higher quality as soon as feed resources expand.

Operation Flood officially aims at the large scale introduction of both cross breds and improved indigenous buffaloes. The difference between the two methods of enhancing productivity is a question of time. The latter process is much slower than the former. Most impact studies hardly mention the crossbred cows even though the envisaged strategy to enhance production is primarily through cross breeding. One study which did mention it [*Chakravarty and Reddy, 1982*] pointed to the limited significance of these animals. Also, the available evidence does not point to a large additional diversion of crop land to fodder crops. The relatively low remuneration from dairy farming has apparently prevented the emergence of capital-intensive dairy farming on a large scale.

In discussing milk production, a distinction should be made between the total production of milk in a region or village and the average (daily) yield of milch animals. In reviewing a number of case studies, van Dorsten [*1990*] arrives at rather mixed results. The available evidence points to an increase in total milk production in certain tracts in Kheda district, a modest production increase in the cooperative villages of Mehsana, and a significant increase in both the Delhi milkshed area (especially the region not covered by the organised market outlet) and Bikaner district. At the same time the evidence does not point convincingly to an increase in the average yield of milch animals, except perhaps in selected tracts of Kheda district and Bikaner villages.

Basically, milk production is a function of the population and productivity of milch animals. But these, in turn, are influenced by various other factors: the need for draught power in agriculture, the size of landholdings, cropping patterns, the level of mechanisation, the demand for milk and dairy products, the commercialisation of milk production, the availability of feed and fodder, the introduction of exotic breeds or upgrading of indigenous ones, and the profitability of milk production. With so many variables, a simple cause–effect relationship between a given production trend and the factors responsible for it is difficult to isolate. A case in point is Tamil Nadu. Dhas [*1990*] establishes that milk production has more than doubled there in 16 years (1966–82). He attributes this rise to increases in both the population and productivity of milch animals but both components increase at a differential rate and the contribution of cows and buffaloes is different (cows contribute more to population and buffaloes to productivity). Dhas does not credit Operation Flood with being the main impetus behind the production increase in Tamil Nadu because it was not designed, or was not able, to affect other factors integral to raising production, such as draught animal needs in agriculture, landholding and cropping patterns, and feed/fodder availability.

THE FOREIGN AID CONNECTION

Foreign aid has had an impact on aspects of India's dairy sector such as the indigenous production base, the incentive to produce and the dependency it may cause. It is useful to explore the motives for dairy aid from both ends as well as to consider how it is used and its implications. India originally viewed food aid as an opportunity to improve her balance of payments situation (at the time of OF's implementation in 1970 India was importing dairy products) as well as to strengthen her dairy sector with a view to achieving self-sufficiency. The NDDB favoured the aid since it could be used to further its model of dairy development.

For foreign donors aid can be regarded as an investment, a means of surplus disposal, or an instrument of development. In the case of technical aid, the investment motive is often preponderant. By providing technical inputs and/or expertise to a recipient country, the donor can expect to benefit later in the form of a secure export market or some other spin-off that would reward its goodwill. As Chatterjee [1990] has shown, this was the case for New Zealand which, during the 1950s, benefited from supplying a steadily increasing share of milk powder exports to India as a reward for technical and human resources which were provided in the first instance as a display of good will. In most cases this type of arrangement implies that inputs for a given project, such as skim milk powder or cross-breeding technology, will be exclusively purchased from the donor. Thus aid usually represents a convergence of interests between donor and recipient agencies. In the case of EC commodity aid to Operation Flood, while the original motives included disposal of surpluses and export promotion, interest subsequently shifted to the need for indigenous dairy development in India to the extent that the EC itself was reluctant to continue the high level of aid. On the other hand, Operation Flood became for the EC a welcome demonstration of the applicability of dairy aid in promoting development, a politically expedient case of 'good surplus disposal'.

Apart from the financial resources which it represents, in India dairy aid seems mainly to have been used to support the national milk grid which aims to keep supply constant in conditions where regional variations and seasonal fluctuations occur. Seasonal variations in milk supply, and the weakness of the production base in some areas, perpetuate the problem of underutilisation of processing capacity. The problem is a vicious circle. As long as a dairy cannot utilise the installed capacity it loses its competitive edge due to high fixed costs per unit of production. It thereby loses its market share in selling milk and is subsequently less able to procure enough milk to achieve efficiency. Dairy commodity aid can be used in such a situation to supplement milk supply at subsidised rates and to contribute to the financial viability of the dairy while at the same time providing funds for investment. Dairy aid from the EC gave the Operation Flood II authorities access to high levels of physical as well as financial resources. The IDC in particular was able to accumulate enormous financial reserves due to the sale of dairy commodities from the EC.

Now, seasonality can be partly avoided by the introduction of crossbred cows which are not seasonal breeders. Regional supply can be evened out by using flush season surpluses in the form of milk powder to cover lean season shortages, or by a transfer of milk from surplus to deficit regions. These, however, are expensive solutions, in particular the first one. While emerging indigenous surpluses in several Indian states would seem to call for the use henceforth of Indian stocks for this purpose rather than foreign ones, the difficulty of adjusting the political economy of milk prices in such a direction may lead to a slow transition. Imported commodities are used to subsidise or substitute for the cost of maintaining an even supply of milk based on domestic resources; they keep the grid running smoothly and relatively inexpensively. In the absence of dairy aid, large amounts of financial resources would be cut off and indigenous supply and demand would have to be readjusted (a potentially painful process, so far averted). Nevertheless, there are signs of hope. Baviskar and Terhal [1990] report that Calcutta is receiving milk powder from surplus areas in India thereby creating interdependency within India rather than external dependence. Moreover, the amount of dairy commodity aid under OF III is substantially less than under OF II. This is in keeping with a more general reorientation of EC policy since the mid-1980s. Commenting on this, Clay [1985: 44] observed that 'perhaps the most striking change is ... the deliberate cutback in dairy aid. This move, reflecting management considerations and widespread concern about the developmental and institutional effectiveness of the uses of these commodities, initiates some degree of autonomy from agricultural surplus disposal in food aid policy-making within the EEC.'

What have been the implications of dairy food aid in the case of Operation Flood? First of all, without access to large amounts of donated skim milk powder and butter oil, Operation Flood would neither have been designed, nor, *a fortiori*, implemented on its present ambitious scale. Aid has been vital to the programme. In particular, NDDB has effectively exploited through Operation Flood the twin advantages of dairy commodity aid: that it enables financial investment in facilities for collecting, processing and marketing indigenous milk; and that it enables the smoothing out of local imbalances in the indigenous marketing process by distribution of the donated commodities themselves.

But the easy access to enormous amounts of dairy aid has also obscured the precarious economic viability of this type of processing and marketing system. The aid flowed partly as a subsidy to the urban milk consumers, while on the other hand giving distant producers access to a new marketing channel. Hence gains were realised in both sectors. But these gains were not soundly enough based on an indigenous cost-reducing breakthrough in productivity. For the latter to occur, higher producer prices would certainly have been necessary, which – given the downward political pressure on consumer prices – would have required even larger amounts of subsidy or aid. Clearly in the case of milk, the price which consumers would have to pay to cover the full costs of a year-round supply would be prohibitively high for poor slum dwellers. Even with subsidies, the present system favours mainly the middle and upper classes. And, while aid has certainly helped many producers by facilitating enlarged sales, here also the favoured regions (for example, Gujarat, Punjab, Maharashtra) have benefited far more than the others.

ACCESS AND SOCIAL BENEFITS

At the beginning of Operation Flood, its sponsors had high expectations of realising social benefits through dairying alongside economic ones. The first charter in particular (Operation Flood I) showed a concern for poverty alleviation. Among the possible effects envisaged were income and employment generation; the ability to overcome social inequities found in class, caste, power and gender constructs; the expansion of a labour intensive pursuit; an equal chance of participation regardless of economic position; scale neutral benefits in that they would accrue equally among participants, not dependent on size of operation; and equitable distribution of benefits across regions. Operation Flood was thought potentially able to achieve reform as its effects permeated society, without eliciting direct confrontation between conflicting interests [*Baviskar, 1990*].

To evaluate OF's record on the distribution of benefits we consider first the evidence pertaining to weaker households. Do landless labourers and marginal farmers benefit to the same degree as other participants? In answering affirmatively, NDDB pointed out that OF I recruited 50–70 per cent of its co-operative members from small and marginal farmers and ten per cent from the landless. First one would wish to distinguish between 'small and marginal' farmers. Secondly, the landless category is an ambiguous one as Baviskar has shown. In his case study of Sanjaya in Kheda district, he found that sons cultivating land registered in their fathers' name, as well as sharecroppers and those with village or town jobs (teachers, small businessmen), are inflating the 'landless labourers' category, although they have access to land and/or fodder. Only one third (33 per cent) of the landless milk producers in Sanjaya are genuine landless labourers; this decreases the official percentage of landless labourer households who keep buffaloes from 30 per cent to 10 per cent. Verhagen [*1990*] further asks, what *proportion* of lower income households holds milch animals, what is the average number of milch animals per household, and what is their quality? His findings similarly indicate that the lowest deciles of the income scale (bottom 10 per cent and 10–20 per cent) are under-represented in terms of milch animal ownership as well as in their rate of participation in dairying. Upper groups own more animals and participate much more in dairying. Even in Kheda district, the homeland of the co-operatives, landless, low-caste, agricultural labourer households have difficulty participating in milk production and sales, according to two studies by Shah [*1981*] and Patel [*1982*]. Verhagen also confirms a positive correlation between landholding and quality of animals kept.

Studies based on a static comparison between villages with and without cooperatives generally concluded that per capita availability of milk among milk producing households in the cooperative villages was larger than in the control villages [*Doornbos et al., 1990: Ch.7*]. Such studies, however, either failed to take into account the consumption of the by-products of *ghee* (clarified butter) manufacture, for example, *khoa* (buttermilk), which is of much greater significance in the control villages, or they suffered from other inconsistencies. Two conclusions clearly emerge from the available evidence. First, in several regions, including those with a longer history of organised milk marketing such as Gujarat and the Delhi milkshed area, weaker section households generally retain very small quantities of milk. This suggests that for the poor producers in these areas milk is indeed a luxury item which they prefer to sell in exchange for cash which can purchase cheaper foodstuffs. Secondly, in view of the probably positive effect of dairy co-operatives on fluid milk sales and the concomitant decline of *ghee* manufacture (and *khoa* consumption) in the villages where they operate, it is improbable that non-producers will have better access to milk and milk products as a consequence of their operations. Although the first phase of OF emphasised the need to distribute subsidised double toned milk among vulnerable groups, this aspect of the programme atrophied.

Whether the programme has significantly increased the availability and hence the consumption of milk and/or dairy products in India, therefore seems doubtful. All studies concur that the establishment of dairy cooperatives has probably led to an increase in fluid milk sales by milk producing households at the expense of other uses, and the little available *trend* evidence on milk utilisation by producer households suggests that it is their *ghee* manufacture that has declined rather than their fluid milk consumption [*Doornbos et al., 1990: Ch.7*].

Verhagen, van Dorsten and Baviskar agree that the poor *have* benefited to some degree from dairying but the problem lies in the increasing gap in benefits between marginal and large producers. As Baviskar notes, among big farmers dairying adds to their profits whereas among small ones it can contribute to survival. Savara [*1990*] shows that this is also true among tribals in Surat (Gujarat) who, despite the fact that the dairy is barely viable, still participate as there is little else to support them economically and dairying is subsidised through the state. Soft loans are also made available to purchase milch animals. As an economic pursuit however, it is not able to sustain itself if the subsidy is withdrawn.

Small farmers, who are a step above those we have defined as the poor, are generally able to benefit rather more from co-operative dairying. The various data on dairy income by landholding category as presented by Patel and Pandey [*1976*], Patel, Thakur and Pandey [*1975*] and Singh and Das [*1982*] show that the average dairy income increases with the size of landholding. An intriguing finding of the 'with and without' studies was that, except in the Sabarkantha villages studied by Singh and Das [*1982*], landless households in the control villages were better off in terms of their average total annual income than their counterparts in the cooperative villages, on several occasions even in terms of their average annual dairy income. The general conclusion is, therefore, that although dairy cooperatives may have increased the earning opportunities of landed milk producers, they certainly have not reduced the general inequalities inherent in the local socio-economic structures. The evidence further suggests that whatever income gains may have been derived by marginal and small producers, have hardly made a significant dent in the major problems of poverty and malnutrition.

Access to land enables fodder to be grown, diversity in fodder production, and higher productivity in the animals. For intensified dairying, the availability of land is an essential requirement. As a result, dairying may in the long run create much the same kind of differentiation as the Green Revolution, in which an overall benefit was observed for many but the rate of benefit was strongly biased in favour of larger farmers who had a stronger land base and other assets or advantages. While Operation Flood authorities stressed that animal ownership was less biased than land ownership, the resource base needed to maximise benefit from milch animals leads one back to the problem of unequal land distribution. The landless and marginal farmer groups are further faced with the difficulty that communal grazing lands are being lost to privatisation, either by land grabbing or by questionable methods of land reform [*Parisot, 1990*]. Bringing the land into co-operative ownership still excludes those outside the co-operatives, so that the best method of preserving access to land is to protect its communality.

In addition to the land constraint on the poor, capital is also a limiting factor. The dairy co-operatives have avoided extending credit to producers, fearing it would induce corruption and preferring to leave credit to the specialised institutions. Medium and large farmers who know how to manage the formal credit system have access to capital, while the poor generally do not. The Integrated Rural Development Programme (IRDP) and other anti-poverty credit schemes have made an effort to alleviate the rural credit situation but quantitative and qualitative limitations have diminished their impact. This leaves many poor producers susceptible to continued relations with private traders who advance them credit and working capital. With the surplus that larger milk producers are able to earn (due to land ownership and more productive animals) they may become able to reinvest and in this way sustain dairying as a profitable enterprise. For others it remains basically unremunerative because of resource constraints (land, capital and quality of animals) and macro-economic policies such as intervention by the central government to keep the procurement price of milk low.

Aside from specific reference to the economically weaker households, income and employment benefits from dairying can be seen in a wider context. In areas where co-operatives have performed reasonably well, milk producers have been able to increase their cash income by increasing the volume of fluid milk sold and by gradually augmenting their bovine holdings. Also, indirect income benefits may have been derived from the services rendered by co-operatives, especially veterinary aid. The discussion of dairy development trends in Mehsana and Bikaner districts shows, however, that these gains may have been partly offset by adverse trends in the cost-price ratio in rural milk production [*Doornbos et al., 1990: Ch.7*]. Moreover, case studies in Khadgodhra village in Kheda district and Anantpur district in Andhra Pradesh revealed that some respondents denied any appreciable income gains. It should be noted that, especially in the case of cattle loan repayments, net income gains are low [*Doornbos et al., 1990: Ch.7*].

Evidence on the employment effects of dairy co-operatives at the household level is contradictory. On the one hand the percentage of workers involved in dairying in the co-operative villages of Mehsana did not register an increase over time [*ibid.*]. Also, in Bikaner district between 1966 and 1976 the relative importance of labour as a cost component of milk production significantly decreased. Both these findings apparently support Shah and Bhargava's [*1982*] hypothesis that dairy co-operatives are labour-saving. On the other hand, except for Shah and Bhargava's analysis, the same studies of Mehsana and Bikaner as well as the study of three OF I milksheds by Singh and Das [*1982*] point to higher levels of employment in dairying in the co-operative, compared to non-co-operative, villages. The only tentative explanation is that the differences in levels of employment observed between the co-operative and control villages existed before the dairy co-operatives were established. Dairy co-operatives themselves, however, did not further increase employment at the household level because their operations had both labour-saving and labour-increasing effects.

PARTICIPATION IN CO-OPERATIVE DAIRYING

Co-operative dairying was officially portrayed as a potential means of overcoming caste, class and gender barriers. Empirical research on these aspects again provides a less optimistic picture. For example, Mitra's work in Bihar [*1990a*] looks at the effect of the Anand pattern in caste/class relations. Corruption in the co-operative in which members acted as middlemen, buying milk from non-members at a low price and selling it at a high price to the co-operative, negated the entire objective of the co-operatives, which was to eliminate exploitation. Mitra observes that although small and marginal producers have access to and use services and inputs from the co-operative dairy, the main benefit to them is the constant price offered by the co-operative for milk.

Old patterns of inequality and exploitation are replicated by Operation Flood in Bihar (and other regions) and this has a negative effect on participatory ideals. Dairy co-operatives do not provide an opportunity for lower classes/castes to catch up with more privileged groups because relatively strong groups can manipulate the co-operatives to their advantage. For example, the permitted practice of awarding bonuses from profits to members based on the quantity of milk sold to the co-operative favours the big producers. A more equitable distribution would be simply to increase the value of the dividends held by all co-operative members [*George, 1990*].

The few studies which investigate the composition of the management committees of village dairy co-operatives indicate that the larger farmer/upper-caste members control the village co-operatives [*Patel, 1982; Chakravarty and Reddy, 1982*]. According to Shah [*1981*] the chairmen and secretaries of Kheda co-operatives are generally from the local dominant castes, which corresponds with the findings of the NDDB study on the socio-economic background of members of the managing committees of Sabarkantha and Rohtak co-operatives [*NDDB, 1977: Exhibit 40–51*]. Moreover, studies which probe this aspect further reveal factional divisions which affect the functioning of the co-operatives [for example, *Rajaram, 1983; NDDB, 1977; Jain, Prasad and Gupta, 1982*]. In fact, even the village Ode described in such positive terms by Somjee and Somjee [*1974, 1978*] is an example of divisive factionalism. Administrative employees, thought to wield more power than elected local officials, usually come from the same caste and class as large farmers. It is not unusual for locally powerful individuals to become co-operative secretaries.

Women's co-operatives have also not been able to avoid caste and class as well as gender inequalities [*Mitra, 1990a; 1990b*]. Baviskar [*1990*] describes the co-operative dairy structure as paternalistic, where managers act and speak on behalf of members in the belief they know better the appropriate course of action. Manager domination militates against member participation. In view of these research findings, one may conclude that dairy co-operatives are neither immune to factional divisions and conflicts nor do they appear to be particularly suited to overcoming the barriers of caste and class.

Further, the only study that attempts to verify empirically the assumption that co-operative dairying is positively correlated with the adoption of a broad range of 'modern' values and practices (thereby making allowance for the influence of interfering variables) could not do so [*Shah, 1981*]. In fact, a negative correlation appeared to exist. In sum, therefore, there is not much empirical evidence supporting the claim that co-operatives act as powerful instruments of change in the direction of a more equitable, less exploitative, rural social structure.

Overall it appears that Operation Flood's main priority was to develop a sophisticated, effective procurement system to satisfy rising urban demand from rural production. The co-operative structure facilitated this process but it also lent itself to a certain kind of rhetoric about poverty alleviation and enhanced social consciousness. This does not imply that poor milk producers have not been helped. With hindsight, however, the reduction of poverty – even while confessed as a goal in official documents on Operation Flood – has never functioned as a decisive criterion for resource allocation and project design. In the absence of a specific agenda to aid the poor, which will require bold changes in power and ownership patterns, existing social gaps are expected to remain, or possibly to be deepened by projects mainly concerned with economic growth.

LONGER-TERM SIDE EFFECTS

Operation Flood had as one of its main objectives an increase in milk production by intensified dairying. However, intensified dairying has certain side effects through the diversion of crop land to fodder crops, the impact on the supply of animal draught power and the disappearance of informal mechanisms of exchange and redistribution. In spite of the dire consequences of these hypothetical long-term effects advanced by some researchers, the focus of most *empirical* studies has generally been on the short-term economic impact or the social implications of co-operative development. Hence there is a lack of definitive data on the long-term 'side' issues.

These implications have become more pronounced with the introduction of cross-breeding with high yielding exotic cattle which was the prime method intended, under Operation Flood to strengthen the production base. The crucial issue is that the male progeny of crossbred milch cattle is less suited for draught and general work. The question, therefore, has arisen whether such a milk production strategy would be detrimental to the rest of the agricultural sector. Moreover, to realise their high milk yield potential cross bred cows need considerably better feeding. As Nair and Dhas point out, this strategy for increasing milk production comes from countries which are able to provide manufactured cattle feed and cultivated green fodder on a large scale, whereas in India the resource base for livestock must be carefully balanced with the needs of the overall agrarian economy and the human population.

INDIA: DAIRY AID AND DEVELOPMENT

The state of milk production in India and the success of incorporating exotic breeds into the indigenous herd display wide variations when disaggregated regionally. Nationally, the evidence is not strong enough to establish the overall status of these important trends. Significant inroads in cross-breeding may lead to serious long-term implications. If crossbreds should come to constitute an increasing proportion of the total cattle population, the preference for female crossbreds may reduce the availability of indigenous males for traction, given a fairly stable total population. Wider trends, such as the commercialisation of dairying, the availability of mechanical power to substitute for draught animals, and the level of remuneration for dairying versus agriculture, will influence the composition of the bovine population as between milch and work animals.

To increase the availability of green fodder from 214.5 million tons in 1973 to 575 million tons by the year 2000 requires more than doubling the area under fodder crops from 6.91 to 16.5 million hectares. Even then it will fall short of estimated requirements at that time by 19.8 million tons (abridged NCA Report [*1976*], cited by Parisot [*1990*]). Turning cultivated land over to fodder production will, however, create problems for food production. While the animal population has been increasing, grazing land has not. Over-stocking degrades the land and further diminishes fodder resources; while neglecting the need for grazing land in order to preserve agriculture, creates a situation where forced migration of livestock will degrade even greater areas of land [*Parisot, 1990*].

If it is true, however, that rural milk producers display a reluctance to allocate additional land to fodder, then the fear of a large scale diversion of food crop land to fodder seems unwarranted. The same reasoning applies to the impact of cross breeding on draught power supply, since the majority of farmers as yet appear to prefer indigenous animals, and for dairying the animal of preference is the buffalo. On the other hand, a decreased draught power orientation of the bovine stock, as indicated by Shah and Bhargava, is serious enough to warrant further attention.

Finally, a different, less quantifiable, side effect of intensified dairying is the decline of 'age old sentiments of mutual co-operation, neighbourly feelings and help to the needy and poor' [*Jain, Prasad and Gupta, 1982: 139*]. In this context one may also mention the custom before Operation Flood of converting milk into *ghee* and generally distributing the *khoa* free to the poor. As the co-operative is mainly concerned with procuring liquid milk, *ghee* production and the disposal of *khoa* has declined. The decline in 'traditional values' should be related to the commercialisation of agriculture in general rather than attributed to the advent of dairy co-operatives as such. But Anand pattern co-operatives, with their commodity approach and exclusion of non-cattle owning households (the poorest of the poor), hardly qualify as the most obvious institutions to correct the social imbalances which result from this process.

FINAL OBSERVATIONS

Co-operative development implies a sustained process on a relatively long-term basis if grassroots participation is to be adequately established. The Operation Flood case tends to confirm that a replication programme of dairy co-operatives which is engineered in a pronouncedly centralised fashion and at relatively high speed is almost bound to become problematic. It is worth recalling in this connection that the Jha Report [*GOI, 1984*] advocated 'pre-co-operative work', whose purpose would be to create the conditions for sustainable grassroots mobilisation. Such participation is the mainstay of a co-operative model of development. The current pattern of expenditure under Operation Flood, which has tended to neglect grassroots extension and education work including the inputs programme component, suggests that the social impetus needed to maintain the human infrastructure of the co-operative has been under-emphasised. To express it in currently fashionable language: too little human capital formation, too much physical capital investment.

Closely related to this point is the sharpened appreciation that a technological revolution in milk-processing, such as Operation Flood exemplifies, is really a different matter from a revolutionary change in the conditions of milk *production.* Here again, long-term trends within a large and complex sub-sector of agriculture are shaped not only by new market opportunities, as stressed by the Operation Flood authorities, but primarily by a number of basic features of farming systems, production factors and ecological conditions. The ways in which project interventions can add to or alter these are by definition limited, and/or take time. In development planning as well as praxis, this is often insufficiently heeded.

On a different level another 'lesson' of Operation Flood would seem to be that too much foreign aid given on easy terms and with little critical monitoring by the donor, tends to reinforce some of the problems just alluded to. Donors such as the EC therefore certainly carry a co-responsibility for a lack of achievement. The EC in particular has accepted and supported the programme too much on its face value.

As the European Court of Auditors observed, '... the Commission referred too often to over-optimistic Indian documents instead of making its own assessments' [*Court of Auditors of the EC, 1988: 13–14*].

By the same token, part of the explanation of and the responsibility for the unprecedented position of influence which the NDDB/IDC have been able to attain within the Indian political-institutional complex should be attributed to the EC. Centralisation in Indian dairy development planning has been salient. To all intents and purposes the NDDB/IDC enjoy a remarkably autonomous position: non-representative, non-governmental, yet vested with vast policy-making powers to intervene in some of the most vital aspects of India's agricultural sector (that is, livestock and land use, dairy technology choice, and institutional articulation between rural and urban sectors with the dairy board acting as a further conduit to the central government and international agencies).

The pivotal position which these key institutions occupy between the 'metropolitan' donor agencies and the various state governments and federations, offers them important scope for manoeuvre and opportunity for brokerage which in turn enhance their autonomous powers. A major impetus in arriving at the pyramidal shape of the Operation Flood structure has been the EC connection and the consequent need for a single channel to receive, distribute and generate capital from donated commodities. This connection has also strongly affected the *direction* of Indian dairy development policy, which might otherwise have been more clearly oriented towards mobilising internal resources, with an eye on the variety of production systems within India. As such, foreign aid has also become a stumbling block to decentralised planning for livestock and dairying in India.

The direction of dairy development promulgated by the NDDB/IDC and the ambitious scale of its goals lead to some basic questions. A leading question is whether the magnitude of operations, and the standardisation, subordination and marginalisation that tend to go with an internationally linked food system (be it through production, processing or distribution), may in the end entail hazards for nutritional self-sufficiency and survival as well as for productive employment. Further questions concern the replicability of projects that have been successful in a particular context; whether projects can be designed to ensure equity of participation and benefits; and whether projects and programmes created with external support can become self-sustaining.

Institutionally, tendencies toward standardisation and centralisation are manifest in several forms. One is the nation-wide application of the Anand pattern under disparate conditions. Another is the integration of the resource base, producers and technology for marketing/processing into a co-operative framework that in the end is managed by a relatively autonomous board. The accumulation of resources in a single body, the IDC (now merged with the NDDB), allowed considerable leverage in influencing policy through investment decisions. As the bulk of investment went into processing and marketing functions rather than enhancing production, this also had implications in terms of who benefited most. Relations with state governments reveal the weight of the NDDB/IDC in negotiations. Because the programme consistently projects such high goals, the normal pace of evolution in livestock improvement and grassroot co-operative formation has been accelerated by top-down interventions to encourage a faster transition.

Equity of opportunity (and this is yet another lesson underlined by the Operation Flood experience) will require unequal public treatment of the less endowed, less powerful and more vulnerable sections of the rural population. Here lies the main need for alternatives to Operation Flood, namely greater dairying opportunities for the poor. These alternatives can be sought in the area of rural production (looking to improved rural nutrition and various processing, transport and marketing techniques), in the area of consumption (low-cost, protein-rich substitutes for milk), and finally in new institutions, such as various other types of co-operative and less centralised dairy development agencies.

REFERENCES

Batra, S.M., 1990, 'Operation Flood: Impact on the Delhi Milk Market', in M. Doornbos and K.N. Nair (eds.).

Baviskar, B.S., 1990, 'Dairy Cooperatives and Rural Development in Gujarat', in M. Doornbos and K.N. Nair (eds.).

Baviskar, B.S. and P. Terhal, 1990, 'Internal Constraints and External Dependence: EEC and Operation Flood', in M. Doornbos and K.N. Nair (eds.).

Chakravarty, T.K. and C.O. Reddy, 1982, 'Dairy Development Programme: Process and Impact: A Study at Village Level in Anantpur, *Journal of Rural Development*, Vol.1 No.14 pp.459–512.

Chatterjee, S., 1990, 'Aid, Trade and Rural Development: A Review of New Zealand's Assistance to Indian Dairying', in M. Doornbos and K.N. Nair (eds.).

Clay, E.J., 1985, 'Review of Food Aid Policy Changes since 1978', *Occasional Papers*, No.1, Rome: World Food Programme.

Court of Auditors of the European Communities, 1988, Special Report No.6/87 on Food Aid Supplied to India between 1978–1985 (Operation Flood II), *Official Journal of the European Communities*, No.C31 (Feb.).

Dhas, A.C., 1990, 'Structure of Milk Production in Tamil Nadu: An Analysis of Trends and Sources of Growth', in M. Doornbos and K.N. Nair (eds.).

Doornbos, M., F. van Dorsten, M. Mitra, and P. Terhal, 1990, *Dairy Aid and*

Development: India's Operation Flood, New Delhi: Sage.
Doornbos, M. and K.N. Nair, 1990, *Resources, Institutions and Strategies: Operation Flood and Indian Dairying*, New Delhi: Sage.
van Dorsten, F., 1990, 'The Impact of Amul on the Milk Economy of Kheda District', in M. Doornbos and K.N. Nair (eds.).
George, S., 1990, 'Operation Flood and Centralised Dairy Development in India', in M. Doornbos and K.N. Nair (eds.).
Government of India, Ministry of Agriculture, 1984, *Report of the Evaluation Committee on Operation Flood II* ('Jha Report'), New Delhi: GOI Press.
Jain, J.L., Prasad, A. and G.N. Gupta, 1982, *Organised Milk Marketing in India, Socio-Economic Impact* (A Case Study of the Delhi Milk Market Scheme in N.W. Rajasthan), Jaipur: Kumarappa Institute of Gram Swaraj.
Lipton, M., 1985, 'Indian Agricultural Development and African Food Strategies. A Role for EC?', in W.M. Callewaert and R. Kumar (eds.) *India: Towards a Common Perspective*, Leuven: Peeters-Leuven, pp. 45–85.
Mitra, M., 1990a, 'Anand in Bihar: Replication of Caste and Class Patterns', in M. Doornbos and K.N. Nair (eds.).
Mitra, M., 1990b, 'Profiles of Women Producers in Andhra Pradesh', in M. Doornbos and K.N. Nair (eds.).
Nair, K.N. and A.C. Dhas, 1990, 'Cattle Breeding Technology and Draught Power Availability: An Unresolved Contradiction', in M. Doornbos and K.N. Nair (eds.).
National Dairy Development Board, 1977, *Pilot Project on Comparative Study of Milk Producers' Socio-Economics, Productivity and Production Practices*, Anand. NDDB.
National Dairy Development Board, 1983, *Dairy India 1983: Operation Flood II*. Anand.
Parisot, R., 1990, 'Cattle Development, Nutritional Requirements and Environmental Implications', in M. Doornbos and K.N. Nair (eds.).
Patel, R.M., 1982, *Glimpses of Change and Development in Borsad Taluka (Kheda district, Gujarat)*, Anand: Sardar Patel University, Agro-Economic Research Centre, Vallabh Vidhyanagar.
Patel, S.M. and M.K. Pandey, 1976, 'Economic Impacts of Kaira District Cooperative Milk Producers' Union (Amul Dairy) in Rural Areas of Kaira District (Gujarat State)', Ahmedabad: Institute of Cooperative Management.
Patel, S.M., D.S. Thakur and M.K. Pandey, 1975, *Impact of Milk Cooperatives in Gujarat*, Ahmedabad: Institute of Cooperative Management.
Rajaram, N., 1983, 'The Structural Linkages of a Crisis in a Milk Cooperative in Kheda District, Gujarat', paper presented at the *Workshop on Cooperatives and Rural Development*, Delhi University, March.
Savara, M., 1990, 'Dairy Development amongst the Tribals in Surat District', in M. Doornbos and K.N. Nair (eds.).
Shah, T. and M. Bhargava, 1982, 'Impact of India's Dairy Cooperatives: Analysis Based on an In-depth Study of Six Villages in the Districts of Sabarkantha (Gujarat), Periyar (Tamil Nadu) and Bikaner (Rajasthan)' Anand: Institute of Rural Management Anand (IRMA).
Shah, V.P., 1981, *Role of Milk Cooperatives in Articulating Rural-Urban Interactions. Experiences in Gujarat, India*, Research Report, Ahmedabad: UNESCO.
Singh, K. and V.M. Das, 1982, *Impact of Operation Flood I at the Village Level*, Monograph No.1, Anand: Institute of Rural Management Anand (IRMA) (July).
Somjee, A.H. and G. Somjee, 1974, 'Cooperative Dairying and the Profiles of Social Change in India', paper presented at the XIX International Dairy Congress, Delhi, Dec.
Somjee, A.H. and G. Somjee, 1978, idem, *Economic Development and Cultural Change*, Vol. 26 No. 3, pp. 577–90.
Terhal, P. and M. Doornbos, 1983, 'Operation Flood: Development and Commercialisation', *Food Policy*, Vol. 8, No. 3.

Verhagen, M., 1990, 'Operation Flood and the Rural Poor', in M. Doornbos and K.N. Nair (eds.).
World Bank, 1978, *India: National Dairy Development Project*, Staff Appraisal Report, Washington, DC.

6

Triangular Transactions, Local Purchases and Exchange Arrangements in Food Aid: A Provisional Review with Special Reference to Sub-Saharan Africa

EDWARD CLAY and CHARLOTTE BENSON

INTRODUCTION

Triangular transactions, local purchases and exchange arrangements have increased significantly in recent years. Donors have entered into these transactions with two sets of concerns in mind: to strengthen food aid operations in terms of increasing cost-effectiveness, speeding up delivery, and providing commodities more appropriate to food habits and customs; and through such transactions, to encourage food production in developing countries and foster and stimulate intra- and inter-country trade.

This chapter aims to provide a brief overview of recent developments involving the various triangular (or trilateral) local purchase and exchange arrangements with special reference to sub-Saharan Africa (SSA), beginning with a discussion of an obvious conceptual problem: the lack of a single phrase to characterise these arrangements which many commentators as well as administrators in donor agencies tend to group together. This awkwardness and ambiguity reflect definitional problems that require some clarification.

DEFINITIONS

The current range of food aid operations that are variously labelled triangular transactions or trilateral food aid transactions, local purchases and exchanges, and swap arrangements, have two features in common. First, under them food aid agencies, including NGOs, have been exercising considerable ingenuity in exploring the possibilities for using food from developing country sources in food aid operations. Second, virtually all these operations involve budgeted financial resources or commodities specifically earmarked for named food aid operations. Where grains are concerned, most of them appear to have been provided as part of the minimum contribution of a donor or (in the case of the EC) group of donors under the Food Aid Convention (FAC).

In recent years, triangular transactions have been the most widely discussed and probably the most important type of operation in SSA involving food aid commodities originating in developing countries. Typically, a donor agency purchases food (with cash from its food aid budget) in one developing source country for shipment to another country, the food aid recipient. Such a transaction is, from the viewpoint of the donor, therefore, an alternative to a normal food aid shipment involving commodities supplied either from a developed exporting country or, to a limited extent, purchased on the international market. Donor countries with commitments under the FAC, but without surplus grain of a suitable type of their own to export in any significant quantity, have been disproportionately important in this type of activity.

Donor agencies also make cash purchases in developing countries for use in food aid actions within the same countries. These operations are normally termed local purchases.[1] Again, by implication, a local purchase is an alternative to a normal food aid operation. Until recently, donors were not able to count such operations as part of their obligations under the FAC. However, this is now allowed and probably explains the greater willingness of some donors to undertake such operations since 1986.

There are, in addition, forms of triangular or trilateral exchange, swap, or barter arrangements which have some of the characteristics of triangular transactions. They combine a normal food aid transaction to a developing country that exports food with a triangular movement of food to a second country.[2] Donors who are grain exporters, and so who are less flexible in being able to use their food aid budgets for the purchase of commodities in third countries, tend to be involved in this type of operation.

There are also commodity exchange or swap operations in which the donor imports food aid commodities into a developing country and then exchanges them for other commodities already in the country, most commonly stocks held by a public food agency, for use in food aid operations in the same country. A study of WFP commodity exchanges has revealed a wide variety of such practices that could be broadly grouped together and may entail either direct barter or the sale and purchase of commodities [*Clay and Benson, 1989*].

More specifically, an imported commodity, most commonly wheat, originating from bilateral donor FAC commitments, may be directly bartered for a second commodity, usually rice in Asia or coarse grains in SSA. Alternatively a donor may provide grain to a food milling organisation and in return take delivery of flour or meal at a fixed exchange rate, normally based on the milling extraction rate. In other cases, like commodities are exchanged, normally on a tonne for tonne basis. This is more common with wheat aid to the South Asian countries – Bangladesh, India and Pakistan – where the final food aid recipients are mostly located at a considerable distance from the point of entry of imported commodities. However, the exchanged commodity may not be of the same type or quality as the one imported. A fourth variant is the importation of commodities for sale and use of the funds realised to purchase locally produced food. Such practices tend to occur in countries where there is no public food agency to market the commodity in question. In this case the distinction between a commodity exchange and monetisation is unclear.

A study by the Relief and Development Institute [*RDI, 1987*] on triangular transactions in SSA pointed to an important and operationally useful distinction between regular and non-regular, or intermittent, developing country exporters of the commodities in question. In those countries which export regularly, such as Thailand and Argentina, the process of purchasing and moving commodities follows normal commercial practice, using established transport routes and generally involving commodities widely traded at the international level. Such purchases are essentially no different from those made in developed countries.

Non-regular exporting countries only intermittently or periodically enter world or regional markets as exporters since they do not consistently have surpluses available. Agencies purchasing from such sources are typically involved in *ad hoc* acquisition and transport arrangements and may have to overcome substantial obstacles. There is also no certainty in such transactions that contracts will be successfully completed. Due to often lengthy negotiating and implementing procedures the country may once again be in deficit before the export contract has been completed. Not surprisingly, there are higher administrative costs associated with exports from non-regular exporting countries.

Much of the debate on the advantages or disadvantages of triangular transactions concerns the non-regular exporting or source economy. However, the classification of countries as regular or non-regular exporters is a question of judgement based on recent trade experience, and there is no general consensus on the set of countries falling into each category.[3]

EVALUATIONS

The fragmentary nature of the literature and the limited number of sources of statistical information on the various operations involving commodities acquired in developing countries reflect the very recent growth of interest in this subject. Three studies have been commissioned by donor agencies to look at the cost-effectiveness, appropriateness and administrative and wider policy implications of these actions. They are the RDI [*1987*] study for the WFP of their triangular transactions and local purchases, the RONCO [*1988*] study of US trilateral operations, and the evaluation commissioned by the Food Studies Group (FSG) of EC triangular transactions [*Hay, 1988*]. A study was also made by Clay and Benson [*1989*] of WFP experience in commodity exchanges (as mentioned above). In addition, a number of articles have discussed the potential advantages and problems associated with such operations [for example, *Shaw, 1983; Jost, 1985*]. In relation to SSA a number of reports and studies have considered these operations in a regional or specific context. Such documentation has largely focused on donor activity, particularly emergency operations, in three regions: the CILSS group of countries; the SADCC countries, particularly Mozambique and Zimbabwe; and Sudan. In addition, a substantial literature of planning and appraisal missions and evaluation reports considers in passing the issues, logistic and administrative, raised by specific operational experiences.

There are two major sources of statistics on these food aid operations, FAO and WFP. FAO began to collect and distribute statistics on triangular transactions, local purchases and triangular swaps from donors in 1982/83. The lists are incomplete but represent the only attempt that has been made to collate different donor purchasing actions. FAO and WFP have made a point of publishing statistics on such operations since 1987 [*FAO, 1987; WFP, 1988a*]. With the growing scale of resources allocated to such operations, donors are also increasingly reporting their food aid purchase and exchange activities separately from other food aid as, for example, the WFP [*1988b*] and Club du Sahel [*1988*]. It is becoming possible, therefore, to document the evolution of these operations with reasonable accuracy at least since 1983/84 generally and for certain donors since the late 1970s [*RDI, 1987*]. The analysis presented in this chapter is based on FAO data for the years 1983/84 to 1987/88, WFP purchase data for the years 1980 to 1986 and WFP exchanges data for the years 1985 to mid-1988. Analysis of exchanges more generally is limited by the scant availability of data.

Earlier studies have confined themselves to the cost-effectiveness issue [*RDI, 1987; RONCO, 1988*]. Most of the policy discussions of these operations have been largely concerned with stating the (potential) opportunity for the cost-effective use of aid resources in this area, as summarised by Shaw [*1983*] and Jost [*1985*]. However, the most recent study has presented the economic issues involved in such operations in a systematic and formal way [*Hay et al., 1988*].

HISTORICAL ORIGINS AND GROWTH

Beginnings: 1963–78

In 1963 the United Nations World Food Programme (WFP) was founded as the first multilateral body concerned exclusively with food aid. It would provide an outlet for donor countries with too small or intermittent surpluses to sustain their own food programme, and would allow richer ones, with resources other than food, a means of supplying food aid. It was also seen as a way of mobilising surpluses in other than donor countries for use as food aid. So from its inception WFP received donations both in cash and kind and some of the cash was spent on purchases on donors' behalf rather than on transport, administration and the other associated costs of food aid. WFP's more recent involvement in purchases in developing countries therefore seemed a natural transition, given its commitments to actively promoting agricultural development and nutritional self-sufficiency and to mobilising surpluses. The first Food Aid Convention in 1967 furthered the opportunity for donors to give cash rather than food by establishing food aid commitments on the part of a number of donors who did not have the export commodities available to meet commitments economically.

However, the amount of food purchased from both developed and developing countries remained very small. For example, between 1963 and 1974 total pledges to the WFP, excluding FAC contributions and miscellaneous income, amounted to US$1,070 million of which 72 per cent was in kind, in the form of commodities. Of the remaining 28 per cent, consisting of cash contributions and services, US$6.2 million was spent on cash purchases. So purchased commodities formed only 0.8 per cent of all those distributed by the WFP – and most of these purchases were of wheat from developed countries.

Two major international initiatives were then taken to encourage cash purchases of food aid in developing countries. First, in 1974, in the wake of the oil crisis of the early 1970s and its subsequent detrimental impact on developing country economies and food production, the World Food Conference adopted an improved policy for food aid. Donor countries were urged to provide food aid programmes with as much additional cash as possible, when appropriate, for commodity purchases from developing countries [*WFP, 1974*]. Second, in 1979, The Guidelines and Criteria for Food Aid, adopted by the Committee on Food Aid Policies and Programmes, recommended that donor countries should provide cash to finance triangular transactions whenever possible and applicable, to diversify the varieties of food provided as aid [*WFP, 1979*].

Until 1977/78, virtually all the WFP's purchases continued to be of wheat, mostly in developed countries. But in that year its acquisitions included 21.9 thousand tonnes of rice and some coarse grains. This marked the beginning of purchases in developing countries on a much wider scale than ever before. At about this time other donor agencies were also experimenting with triangular transactions. For example, the EC funded a shipment of 500 tonnes of white maize from Zambia to Botswana under its 1978 programme [*IDS/CEAS, 1981*]. Then, in 1979, two large-scale operations, the Thai Kampuchean Border Operation and the Zimbabwe Maize Train, both of which involved considerable volumes of triangular transactions, focused attention on the possible positive effects of these operations.

The First Large-Scale Triangular Operations: 1979–83

Thai–Kampuchean Border Relief Operation

The large international refugee relief operation on the Thai–Kampuchean border, which began in 1979, led to an enormous demand for food aid, especially rice, the recipients' staple cereal. Thailand, itself one of the world's major rice exporters, and Burma, also an important one, were therefore the most obvious sources of rice. In 1980, at the height of the relief operation, WFP alone made purchases to the value of US$57.5 million (including some dried fish, canned fish and vegetable oil as well as rice). This constituted 73 per cent of total WFP cash purchases in that year.

However, the large volume of food aid acquired in Thailand for the relief operations was not simply the outcome of a decision by the donor community. Thailand forced the donors' hands by refusing to allow the import of any rice into the country for refugee feeding. Otherwise some purchases would probably have been made in North America, Europe and Australia as well.

Zimbabwe 'Maize Train' Operation

Zimbabwe is a land-locked country, dependent on transport networks in neighbouring states for international trade. The long period of civil strife in the country, coupled with that in Mozambique, Zimbabwe's main natural throughway to sea ports, had left the transport system very run-down by 1980. Zimbabwe had two successive bumper crops of maize (two million tonnes in 1980/81 and 1.5 million tonnes in 1981/82). Simultaneous increases in the production of other export crops placed enormous strains on transport resources. Demand for transport far out stripped supply and the Zimbabwe Government was forced to devise a strict system of priorities in the allocation of wagons for the movement of goods. Maize, as a comparatively high volume, low value, good ranked low on the list of priorities. The Zimbabwean Grain Marketing Board, therefore, was faced with the prospect of having to stockpile maize for two or three years until the transport situation had improved.[4] This would have represented a burden on both its storage and financial resources and it was therefore decided to approach the WFP for assistance in reducing surpluses.

Simultaneously, other parts of Africa were facing acute food shortages; and the WFP was trying to implement triangular transactions on a significant level. The Zimbabwean maize surpluses coupled with deficits elsewhere in SSA offered the perfect opportunity to achieve this. And so, in April 1981, the Zimbabwe 'maize train' operations began.

Severe transport problems were encountered, largely because the maize was being moved through Mozambique, both for use there and for export by sea. Nevertheless, by the end of 1983 over 400,000 tonnes of maize had been delivered to a total of 18 different African countries, for use both in emergency operations and as grant food aid in support of development projects. The operation involved about twenty different donors and was the first co-operative donor effort to support triangular transactions. Donations included expertise, cash and urgently needed equipment with which to improve transportation and logistics facilities as well as cash donations to finance purchase of the maize. A major effort was made to reconstruct the transport and communications infrastructure.

However, the drought affecting many other African countries did not leave Zimbabwe unscathed. In 1983 the quantity of maize moved by the maize train fell; and by 1984 the Government of Zimbabwe had to import maize. Outstanding maize train deliveries were deferred by mutual consent and were finally delivered in 1985, only after another bumper crop. 1986 was also a good year but in 1987 Zimbabwe's worst drought in 40 years severely affected maize production and exports were once again halted. Provisional figures for maize production in 1988 suggest a crop not much smaller than in 1985 and 1986.

The Zimbabwe operations provided an early indication that, even in African countries with more regular surpluses, these cannot be guaranteed. Moreover, donor cooperation of the early 1980s was not repeated after the good 1985 harvest. Since then, individual donors have made separate arrangements with the Zimbabwe Maize Marketing Board and the Board has also made commercial export sales.

The African Food Crisis and its Aftermath: 1983–88

The African food crisis of 1982–86 provided some opportunities for triangular transactions, local purchases and exchange arrangements from a number of source countries in SSA. Drought did not affect all countries, or all regions of certain countries, simultaneously and some food surpluses were available in parts of the continent throughout the period. Meanwhile some donors, despite substantial domestic surpluses, were willing to provide food aid in cash, rather than in kind, to facilitate purchases. There have been many recent resolutions by African regional meetings favouring triangular food aid. For example, the report of the Abidjan Seminar, sponsored by the African Development Bank and WFP, provided further examples of the strong desire to increase the role of triangular transactions.

The African food crisis waned after improved harvests in 1985/86 led to substantial agricultural surpluses in some countries, and in 1986/87 total food aid to SSA was only about 40 per cent of the peak levels of 1984/85. But in some SSA countries food deficits continued, necessitating food imports. For example, a CILSS report [*1988*] noted that harvests in the north of several Sahelian countries in 1987 had been poor while those in the south were quite favourable. An opportunity for local purchases therefore existed and a concerted effort was made by the international donor community to respond. In 1986/87 within

SSA triangular transactions, trilateral exchanges and local purchases increased by 52 per cent over the previous year's level to 414,000 tonnes and in 1987/88 increased by a further 23 per cent to 544,000 tonnes.[5]

During this post-crisis period there has also been some switching between sources, and some problems of implementation reflecting mainly, but not only, problems in source countries and regions. In early 1987 some exporting countries embargoed or severely restricted purchases and shipments for export, even under existing contracts. The governments in Kenya and Zimbabwe had become sensitive to the risks of a rapid reduction in stocks and possible supply problems following a poorer harvest. Other countries, for example Malawi and Niger, were again moving temporarily into deficit. But the feared supply problems did not materialise, except in Malawi. There were good harvests in 1988 and Sahelian states are again looking for an enhanced donor commitment to triangular and local purchase operations.

Commodity Exchanges

Until the 1980s commodity exchanges were not viewed particularly favourably, although their usefulness in supplying a more appropriate commodity to the final recipient was recognised [*Clay and Benson, 1989*]. However, during the 1980s they have emerged as an increasingly important part of donor food aid programmes.[6] Exchanges were used in a number of instances where distribution difficulties made them the most cost-effective means of supply, for example when the distribution point of the food aid was a considerable distance from its point of entry into the country.

Exchanges have also been seen as a useful device for providing food aid to countries which have become self-sufficient in cereal staples but still contain population groups that would benefit from food aid projects. In particular this is true of a number of south and south east Asian countries. Exchanges have also offered the opportunity of providing cereal food aid for countries where the import of some processed commodities has been prohibited to protect the local milling industry.

Between 1986 and 1988 donors also increasingly engaged in both triangular and local exchange arrangements due to budgetary constraints on the level of financial resources available for commodity purchases. However, the basis on which exchange ratios are determined remains contentious since there is no system which is fair to all parties at all times.

TABLE 1
TOTAL TRIANGULAR TRANSACTIONS, TRILATERAL EXCHANGES AND LOCAL PURCHASES OF CEREALS, 1983/84–1987/78, ARRANGED BY CATEGORY OF COUNTRY OF PURCHASE (IN GRAIN EQUIVALENT THOUSAND TONNES)

	1983/84	1984/85	1985/86	1986/87	1987/88
LDC Regular Exporter (a)	199 (b)	462	482	428 (d)	318
LDC Non-Regular Exporter	221	244	200 (c)	330 (e)	453
Local Purchase	23	27	140	169	234
Unknown LDC Source	0	3	9	28	166
Total Purchases	443	736	831	954	1,171
Total Cereals Food Aid	9,850	12,511	10,949	12,579	13,176
Total Purchases as % of Total Cereals Food Aid	4.5%	5.9%	7.6%	7.6%	8.9%

Source: FAO unpublished data.

(a) Countries classified as LDC regular exporters are Argentina, Burma, China, Pakistan and Thailand.
(b) Includes 39,885 tonnes to an unknown destination (so might be a local purchase).
(c) Includes 10,000 tonnes to an unknown destination (so might be a local purchase).
(d) Includes 11,448 tonnes to an unknown destination (so might be a local purchase).

TRENDS AND PATTERNS DURING THE MID-1980s[7]

Overall Trends

The scale of triangular and local purchase operations probably increased rapidly between 1979/80 and 1981/82 but then fell away rapidly due to supply problems in SSA. However, the incomplete data for these years do not permit detailed analysis. Since 1983/84 triangular transactions, trilateral exchanges and local purchases have accounted for an increasing share of food aid operations both globally (Table 1) and in SSA (Table 2). Between 1983/84 and 1987/88, total purchases and trilateral exchanges in developing countries increased from less than half a million tonnes to almost 1.2 million tonnes.[8] As a percentage of total cereals food aid they also gradually increased, from 4.5 per cent in 1983/84 to 8.9 per cent in 1987/88.[9] In addition, some 1.2 million tonnes of cereals were acquired through local exchange by WFP

TABLE 2
TOTAL TRIANGULAR TRANSACTIONS, TRILATERAL EXCHANGES AND LOCAL PURCHASES OF CEREALS RECEIVED BY SUB-SAHARAN AFRICAN COUNTRIES, 1983/84 TO 1987/88 (IN GRAIN EQUIVALENT THOUSAND TONNES)

	1983/84	1984/85	1985/86	1986/87	1987/88
LDC Regular Exporter (a)	55	331	299	263	166
LDC Non-Regular Exporter	213	235 (b)	194 (c)	313 (d)	449 (e)
of which from SSA Countries	193	187 (b)	180 (c)	312 (d)	418 (e)
Local Purchase	15	19	110	128	125
From Unspecified Origin	1	3	3	28	166
Total	283	586	607	732	906
As % of total food aid to SSA	9.4%	11.8%	14.1%	22.6%	24.6%
As % of total purchases	64.0%	79.6%	73.1%	76.7%	77.4%

Source: FAO unpublished data and 'Food Aid in Figures', various, Rome.

(a) Countries classified as LDC regular exporters are Argentina, Burma, China, Pakistan and Thailand and Turkey.
(b) Includes 1,017 tonnes originating in sub-Saharan Africa but for use in an unknown country. Therefore they could be local purchases or triangular transactions destined for either sub-Saharan African or non-sub-Saharan African countries.
(c) Includes 10,000 tonnes originating in sub-Saharan Africa but for use in an unknown country. Therefore they could be local purchases or triangular transactions destined for either sub-Saharan African or non-sub-Saharan African countries.
(d) Includes 14,000 tonnes originating in sub-Saharan Africa but for use in an unknown country. Therefore they could be local purchases or triangular transactions destined for either sub-Saharan African or non-sub-Saharan African countries.
(e) Includes 9,948 tonnes originating in sub-Saharan Africa but for use in an unknown country. Therefore they could be local purchases or triangular transactions destined for either sub-Saharan African or non-sub-Saharan African countries.

between 1985 and mid-1988 (see note 5). By 1987/88 the overall total of all these non-conventional food aid operations had probably risen to around 1.4 to 1.5 million tonnes or ten to 11 per cent of total cereals food aid. What was only a policy aspiration in the 1970s has become a significant part of overall food aid operations.

The relative importance of the different types of purchase has also changed over time. Since 1984/85 there has been a decline in the relative importance of triangular transactions and trilateral exchanges involving regular exporter countries, while the volume of such actions originating in non-regular exporter countries, and of local purchases, have both increased significantly.

Three-quarters of all purchases and trilateral exchanges went to sub-Saharan African countries between 1983/84 and 1987/88. In this region alone the proportion of such actions in total food aid increased from nine per cent in 1983/84 to a quarter in 1987/88. Although a significant proportion of the food originated in developing countries outside sub-Saharan Africa, these gradually declined in importance until in 1987/88 triangular transactions and trilateral exchanges among SSA countries represented almost half of total purchases and trilateral exchanges received in the region.[10]

However, over half of WFP commodities imported for exchange over the period 1985 to mid-1988 went to Asia. This pattern partly reflects the fact that a number of south and south east Asian countries have achieved self-sufficiency in their cereal staple production so that exchanges were particularly appropriate there.

Commodity Composition of Cash Purchases

Total triangular transactions, trilateral exchanges and local purchases by commodity, by category of source of purchase, and as a proportion of total food aid in that commodity, are shown in Table 3.[11] Wheat formed a very insignificant part of purchases although it represented over three-quarters of total cereals food aid between 1983/84 and 1986/87. Rice constituted 43 per cent of all purchases and trilateral exchanges, although it formed only nine per cent of total food aid. A third of total rice food aid was acquired through purchases and trilateral exchanges, 88 per cent of them originating in regular exporting developing countries, particularly Thailand and Burma, for use elsewhere. About three-quarters (76 per cent) of all rice purchases were made by just one donor, Japan, in 1984/85 and 1986/87 (see below under 'Donors').

Coarse grains, which constituted only 15 per cent of total food aid, accounted for over half (53 per cent) of purchases and trilateral exchanges. Some 21 per cent of total coarse grain food aid was acquired this way, of which about three-quarters (73 per cent) represented triangular transactions and trilateral exchanges in non-regular exporter countries, particularly Kenya, Malawi and Zimbabwe, mostly for use in other African countries.

A similar pattern emerges from the WFP data on cereals purchases.

TABLE 3
COMMODITY COMPOSITION OF TRIANGULAR TRANSACTIONS, TRILATERAL EXCHANGES AND LOCAL PURCHASES,
1983/84 TO 1987/88 (IN GRAIN EQUIVALENT THOUSAND TONNES)

COMMODITY	CATEGORY OF COUNTRY OF PURCHASE	1983/4	1984/5	1985/6	1986/7	1987/8(a)	TOTAL
COARSE GRAINS	LDC Regular Exporter	7	90	82	2	39	220
	LDC Non-Regular Exporter	213	240	198	342	603	1,595
	Local Purchases	17	21	107	117	112	374
	TOTAL	237	351	386	461	754	2,189
	as % total coarse grains food aid	18.5%	14.7%	21.3%	34.3%		
RICE	LDC Regular Exporter	191	372	377	372	261	1,575
	LDC Non-Regular Exporter	9	4	6	6	4	29
	Local Purchases	3	9	33	24	119	187
	TOTAL	203	385	417	402	384	1,791
	as % total rice food is	17.9%	38.7%	35.5%	41.7%		
WHEAT/ WHEAT FLOUR	LDC Regular Exporter	3	0	28	59	29	117
	LDC Non-Regular Exporter	0	0	0	5	0	5
	Local Purchases	0	0	0	28	3	31
	TOTAL	3	0	28	91	31	153
	as % of total wheat/wheat flour food aid	( )	.0%	.4%	.9%		
GRAND TOTAL		443	736	831	954	1,169	4,133

Source: FAO, unpublished data and *Food Aid in Figures*, various, Rome.

(a) There was also a purchase of 2,000 tonnes of unspecified cereals from an unknown source country.

This also covers non-cereals, and reveals that pulses are the other commodity type of which purchases in developing countries have been an important means of acquisition. Although they formed under four per cent of total purchases, and only about 1.3 per cent of all food distributed by WFP in 1986, 55 per cent of pulses were acquired in developing countries, as was about 23 per cent of fish (representing 4,700 tonnes), mostly bought from Thailand.

Overall, both FAO and WFP data indicate that triangular purchases and trilateral exchanges involving regular exporting countries have been dominated by rice, particularly from Thailand, while such actions from non-regular exporters have been dominated by coarse grains, particularly for movement within Africa.[12] Both commodities were much more important in total purchases and trilateral exchanges than in total food aid.

The higher proportion of rice and coarse grains than of wheat acquired through purchases and trilateral exchanges is partly an indication that the types of these commodities grown in donor countries are inappropriate for consumption in developing countries. Purchases have largely involved commodities with rather specific culinary and other characteristics related to local and regional markets. In fact, a look at the types of food aid commodities supplied over a longer period reaching back to 1970 shows that coarse grains have only become a significant element since 1980/81 when cash purchases began on a larger scale.

As already discussed, both the FAO and WFP sets of data also indicate a substantial increase in local purchases since 1985/86 (or 1986, depending on the reporting period used). Sorghum from Sudan has featured prominently amongst these. The commodities acquired by the WFP through exchanges indicate that these were aimed at supplying more appropriate food. The most important exchanges were of wheat and wheat flour for rice in Asia. The second most important were of wheat and wheat flour for coarse grains in sub-Saharan Africa.

End Use of Commodities Acquired in Purchase and Exchange Actions

The only systematic information available on the end use of commodities is for WFP operations in 1985 and 1986 [*RDI, 1987*]. These commodities, particularly from non-regular exporting developing countries, were used disproportionately for emergencies rather than as development programme assistance or for projects. In both years at least two-thirds, and sometimes much more, of all WFP purchases from non-regular developing country exporters and of local purchases, on WFP's own behalf and for other donors, were used for emergency actions. During approximately the same period just under half of total food aid to sub-Saharan Africa was for emergency use, about 20 per cent for project aid and the remainder for programme aid. The implications of these findings are explored in greater depth below ('Implications for donor support for food security').

However, although the WFP has entered into a number of *ad hoc* exchanges as and when local surpluses have permitted, for use in both emergency and development projects, a substantial proportion of its exchanges has been planned in the context of development projects. In countries where supply is less certain, exchanges have sometimes been chosen as the means of meeting certain food aid commitments but they have been backed by contingency plans to import the chosen commodities, or acceptable alternatives, directly should they become unavailable locally.

Variability of Non-Regular Sources

An important feature of all triangular and exchange operations has been their variability, with the suppliers of one year quite often becoming recipients in the next (Table 4). The WFP's experience also illustrates the *ad hoc* nature of local purchases, with donors responding to periodic availabilities of local surpluses. From 1980 to 1986 local purchases were made in 38 countries, but in only nine of these was the same commodity bought in three or more years; and in only 11 of them were local purchases of any commodity at all made in more than three years. Countries benefiting from local purchases on a more regular basis were Malawi, Rwanda, Burundi, India and Nepal.

Availability of Surpluses

Differences in the level of production, and hence in continuity of export availability, is the major difference between regular and non-regular exporter sources. The movement in indices of per capita total cereal production in Zimbabwe and Thailand, exporting countries in which the most cash purchases and trilateral exchanges were made in the mid-1980s, graphically contrasts the relative instability in production of a non-regular, compared to that of a regular, exporter (Figure 1).

Transient surpluses, coupled with time lags in negotiating and implementing triangular transactions, can lead to the partial or total cancellation of a purchase. For example, a purchase of 10,000 tonnes of

TABLE 4
TOTAL TRIANGULAR TRANSACTIONS, TRILATERAL EXCHANGES AND LOCAL PURCHASES BY SOURCE,
1983/4 TO 1987/8 (IN GRAIN EQUIVALENT THOUSAND TONNES)

SOURCE COUNTRY	1983/84	1984/85	1985/86	1986/87	1987/88	TOTAL 1983/84 TO 1987/88	AS %
Sub-Saharan Africa							
BENIN	–	5.0	4.0	.4	–	9.4	.23%
BOTSWANA	.5	.2	–	–	–	.6	.02%
BURKINA FASO	.5	–	17.3	4.5	7.8	30.1	.73%
BURUNDI	–	–	–	.3	.4	.7	.02%
CAMEROON	–	4.9	8.8	6.6	3.9	24.1	.59%
CHAD	–	.4	–	4.5	4.3	9.1	.22%
COTE D'IVOIRE	–	1.5	10.6	4.0	–	16.1	.39%
ETHIOPIA	–	–	–	6.0	6.7	12.7	.31%
GAMBIA	–	–	–	1.6	–	1.6	.04%
GHANA	–	15.0	–	–	–	15.0	.36%
GUINEAU BISSAU	–	–	–	–	.1	.1	( )
KENYA	85.0	14.9	10.7	82.6	54.8	247.9	6.03%
LESOTHO	–	–	–	–	4.9	4.9	.12%
MADAGASCAR	–	–	–	–	2.1	2.1	.05%
MALAWI	46.0	101.3	34.8	23.1	8.0	213.2	5.19%

TRIANGULAR TRANSACTIONS, LOCAL PURCHASES, EXCHANGES

MALI	.3	.8	17.9	13.8	23.9	56.7	1.38%
MAURITANIA	—	—	—	3.0	6.5	9.5	.23%
NIGER	14.4	22.5	11.2	7.5	4.4	59.9	1.46%
RWANDA	2.9	2.2	.3	1.1	1.6	8.1	.20%
SENEGAL	—	—	—	5.0	3.1	8.1	.20%
SOMALIA	—	—	1.5	17.1	4.8	23.4	.57%
SUDAN	15.7	.3	45.6	51.9	32.2	145.7	3.54%
SWAZILAND	.2	—	—	—	1.0	1.2	.03%
TANZANIA	—	—	—	—	32.8	32.8	.79%
TOGO	.9	5.8	5.1	—	—	11.7	.29%
UGANDA	—	2.0	3.0	—	1.4	6.4	.16%
ZAIRE	—	—	—	3.3	2.9	6.2	.15%
ZAMBIA	—	—	—	.8	—	.8	.02%
ZIMBABWE	41.7	28.9	119.5	204.2	350.0	744.2	18.11%
AFRICA UNSPECIFIED	—	—	—	5.3	—	5.3	.13%
Total	208.1	205.5	290.1	446.6	557.5	1707.7	41.17%
Mediterranean, Middle East and North Africa							
LEBANON	—	—	—	2.5	—	2.5	.06%
SAUDI ARABIA	—	—	—	20.0	10.5	30.5	.74%
TURKEY	—	—	—	—	6.9	6.9	.17%
Total	—	—	—	22.5	17.4	39.9	.80%

TABLE 4 (continued)
TOTAL TRIANGULAR TRANSACTIONS, TRILATERAL EXCHANGES AND LOCAL PURCHASES BY SOURCE, 1983/4 TO 1987/8 (IN GRAIN EQUIVALENT THOUSAND TONNES)

SOURCE COUNTRY	1983/84	1984/85	1985/86	1986/87	1987/88	TOTAL 1983/84 TO 1987/88	AS %
Asia and Pacific							
BANGLADESH	–	–	–	–	.4	.4	.01%
BHUTAN	–	–	–	1.2	2.0	3.3	.08%
BURMA	58.1	78.0	80.5	95.9[b]	1.6	314.1	7.64%
CHINA	–	62.8	4.7	–	22.3	89.7	2.18%
FIJI	–	–	–	( )	–	( )	.00%
INDIA	–	17.9	4.6	5.8	–	28.3	.69%
INDONESIA	19.5	4.8	.7	1.1	–	26.1	.63%
NEPAL	–	–	–	–	3.6	3.6	.09%
PAKISTAN	23.8	35.3	52.2[a]	118.5	88.0	317.9	7.73%
PHILIPPINES	–	–	–	1.5	–	1.5	.04%
THAILAND	110.3	233.6	336.2	192.2	271.3	1143.6	27.82%
VIETNAM	5.8	2.0	–	–	–	7.8	.19%
Total	217.5	434.4	478.9	416.2	389.1	1936.2	47.10%

TRIANGULAR TRANSACTIONS, LOCAL PURCHASES, EXCHANGES

Central and Latin America

						Total	
ARGENTINA	7.4	56.2	34.3	35.5	18.2	151.6	3.69%
BOLIVIA	–	–	–	.3	1.2	1.4	.03%
COLOMBIA	4.7	1.7	–	–	–	6.4	.11%
COSTA RICA	–	–	–	.5	–	.5	.01%
ECUADOR	–	–	–	–	3.9	3.9	.10%
EL SALVADOR	–	–	2.3	.3	–	2.6	.06%
GUATEMALA	–	5.0	5.0	–	17.0	27.0	.66%
GUYANA	–	–	–	4.4	–	4.4	.11%
HONDURAS	1.6	27.5	11.0	( )	.7	40.8	.99%
MEXICO	–	3.0	–	–	–	3.0	.07%
ST. CHRISTOPHER & NEVIS	–	( )	–	–	–	( )	.00%
ST. LUCIA	–	( )	–	–	–	( )	.00%
SURINAM	–	–	.5	–	–	.5	.01%
Total	13.7	93.4	53.1	41.0	41.0	242.1	5.85%
UNSPECIFIED	3.4	3.0	8.9	27.7	166.0	208.9	5.08%
TOTAL	442.6	736.2	830.9	954.1	1171.1	4134.8	100.00%

Source: Unpublished FAO data.

a. – Includes 14,000 tonnes in cereals equivalent of rice for which source specified is Thailand/Pakistan.
b. – Includes 21,000 tonnes in cereals equivalent of rice for which source specified is Burma/Thailand.

white maize from Malawi for use in Tanzania was made by the EC following Malawi's bumper maize harvest of 1984. However, movement of the maize was delayed until March 1986. Further delays (mainly transport difficulties) meant that delivery was spread over more than a year. By May 1987, 8,587 tonnes had been delivered. However, although Malawi had another good maize crop in 1986, by April 1987 it was clear that that year's harvest was going to be very poor and so the Malawi Government began to place restrictions on the export of maize. Meanwhile Tanzania had excellent maize harvests in 1986 and 1987. Indeed some donors investigated the possibility of purchasing maize in Tanzania for use in Malawi in 1987 but decided against it because of grain borer infestation of the Tanzanian crop [*Clay and Benson, 1988*].

Impact on International Trade

Triangular transactions and trilateral exchanges do not just entail intra-regional movements of food. Figure 2 illustrates the distances involved in implementing the distribution of cereals originating in Zimbabwe in 1986/87 and destined for other SSA countries. Small volumes of pulses have also been transported over great distances within Africa. For example, in 1986 consignments of 560, 50 and 136 tonnes of pulses were purchased by the WFP in Kenya for use in Cape Verde, Togo, and São Tomé and Principe, respectively.

A comparison of total cereal exports from, and triangular transactions originating in, SSA source countries between 1983/84 and 1986/87 suggests that purchases and exchanges have come to represent an important element in intra-African trade (Table 5). The 1988 study by Hay and others calculated that triangular transactions constitute over 40 per cent of intra-regional cereal trade and have increased the volume of inter-regional trade.

However, there are severe data problems concerning intra-African trade, much of which is not recorded. The degree to which commodities are traded informally between neighbouring states is difficult, if not impossible, to measure but in some instances is likely to be substantial. The available statistics are also generally not disaggregated by commodity or by origin or destination. Information on triangular actions is often the only available data on trade between two countries. So it might be more true to say that triangular food aid operations appear to be an important element of recorded trade, especially in east and southern Africa.

The patterns of legal, recorded trade also partly explain why

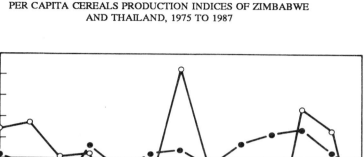

FIGURE 1
PER CAPITA CEREALS PRODUCTION INDICES OF ZIMBABWE
AND THAILAND, 1975 TO 1987

Source: Based on data from FAO, *Production Yearbook*, various.

triangular transactions and trilateral exchanges have frequently encountered management problems. Previously the relevant commodities were officially traded only in small quantities, if at all. Governments are normally more concerned with the (more lucrative) export of commercial crops and accord low priority to the export of staple food crops. Internal food security concerns have also made them reluctant to allow cereal exports except when they experience unanticipated surpluses.

Donors

Some of the smaller food aid donors have been disproportionately important as sources of funding for cash purchases in developing

FIGURE 2
TRIANGULAR TRANSACTIONS OF CEREALS ORIGINATING IN ZIMBABWE AND DESTINED FOR OTHER SSA COUNTRIES, 1986/87

→ direction and volume (in thousand tonnes) of movement

Source: Based on FAO unpublished data.

TABLE 5
COMPARISON OF TRIANGULAR TRANSACTIONS OF CEREALS ORIGINATING IN SUB-SAHARAN AFRICAN COUNTRIES WITH TOTAL CEREALS EXPORTS FROM THOSE COUNTRIES, 1983/84–1987/88
(IN GRAIN EQUIVALENT THOUSAND TONNES)

Country of purchase	TRIANGULAR TRANSACTIONS					TOTAL TRIANGULAR PURCHASES 1983/84–1987/88	RECORDED CEREALS EXPORTS 1984–88	TRIANGULAR TRANSACTIONS AS % EXPORTS
	1983/84	1984/85	1985/86	1986/87	1987/88			
BENIN	–	–	–	–	–	5.0	na	
CAMEROON	–	5.0	2.0	–	1.6	5.8	15.1	38.6%
COTE D'IVOIRE	–	2.3	1.5	10.6	4.0	16.1	99.6	16.2%
GHANA	–	1.5	15.0	–	–	15.0	na	
KENYA	85.0	15.0	12.5	10.7	82.6	54.8	245.6	741.8	33.1%
MALAWI	44.5	98.5	30.5	15.8	–	189.3	365.0	51.9%
MALI	–	–	–	–	7.5	7.5	na	
NIGER	13.3	18.1	–	–	–	31.4	na	
SENEGAL	–	–	–	–	3.0	3.0	na	
SUDAN	13.0	–	–	24.5	7.4	44.9	1145.2	3.9%
SWAZILAND	.2	–	–	–	1.0	1.2	na	
TANZANIA	–	–	–	–	21.0	21.0	na	
TOGO	.9	5.8	4.8	–	–	11.5	12.9	89.2%
UGANDA	–	1.5	–	–	–	1.5	na	
ZIMBABWE	36.2	26.7	117.7	191.4	336.0	707.9	1190.7	59.5%
TOTAL	193.1	187.0	176.2	318.4	432.2	1306.8		

Source: FAO, unpublished data.
na – not available

countries. These are donors with commitments under the FAC but without surplus grain of their own to export in any significant quantity – for example, Japan, the Netherlands, the Federal Republic of Germany and the United Kingdom. Purchases are becoming an increasingly important part of some donor programmes and in 1987/88 represented over 95 per cent of total food aid donated by the Netherlands, Switzerland, Norway and Denmark. In contrast, the larger food aid donors, who have tended to be major developed country cereal exporters, have been less important purchasers.

Japan has been the donor most involved in cash purchases in developing countries, particularly since 1984/85. This one country funded approximately a third of total cash purchases between 1983/84 and 1987/88, although in the same period it only contributed under four per cent of total cereals food aid. In fact, since 1984 Japan has implemented an explicit policy of providing all food aid as rice purchased from developing countries (mainly Thailand, Burma and Pakistan), and intends to make triangular transactions the main vehicle of its cereal food aid in the future. In addition to its rice purchases, Japan has been buying cereals in SSA. All its bilateral food aid contributions are now made through this mechanism.[13]

In contrast, the larger food aid donors, who have tended to be major developed country cereal exporters, have been less important purchasers. The USA, which has contributed approximately 60 per cent of all cereals food aid since the mid-1980s and has always been the largest food aid donor, has been much less important in terms of financing purchases in developing countries. In fact, like some other donors for whom purchases have formed a very small proportion of total food aid, the USA has preferred to become engaged in exchange transactions as a way of supporting Third World exporters such as Zimbabwe. These exchanges have been acceptable to the donors' domestic agricultural lobbies which would possibly have been opposed to purchases when there were already substantial domestic surpluses.

The exception to this negative relationship between the level of developing country purchases in a donor's total food aid commitments and its level of cereal surpluses is the EC's Community Action programme. The EC, with considerable exportable surpluses of grain, was yet the second most important funder of purchases in the mid-1980s. Clearly, a different balance of factors has influenced EC policy. There is strong support from some member states for the EC Commission's positive attitude to purchasing in developing countries, and the Directorate General for Development (DGVIII) has sought to mobilise food aid directly from agricultural markets.

TRIANGULAR TRANSACTIONS, LOCAL PURCHASES, EXCHANGES

ECONOMIC ASSESSMENT

The arguments in favour of triangular transactions and similar food aid operations include:

(i) Creation of additional demand in the exporting country or region. When parastatal agencies build up stocks surplus to current distribution requirements, increased demand in the form of exports and purchases for local food aid operations reduces the danger of a 'policy cycle', in which price reductions and decreased purchases from producers lead to reduced acreage sown the next year. (There is a strong presumption in the economic literature for the individual economy that trade is more cost-effective than inter-year storage as a counter-cyclical measure.) Avoiding a policy cycle is particularly important when bumper crops are largely the result of good weather, and when reduced planting for the next crop may be coupled with much poorer weather. Triangular and similar actions are therefore *prima facie* attractive as measures to foster trade both within and between African economies.

(ii) Promotion of exports with foreign exchange gains to the producing countries. Similarly, trilateral exchange operations can lead to foreign exchange savings when the commodity imported for exchange replaces commercial imports which would otherwise be made.

(iii) Provision of more appropriate food aid commodities, especially for rural populations. There are potentially immediate benefits, in terms of acceptability and ease of processing, in providing more appropriate commodities. This implies a higher value to final recipients. It also reduces longer-term risks of shifting consumer tastes further towards imported commodities which cannot be produced domestically and so must be imported to meet demand, thus posing a potentially severe foreign exchange problem for many lower income countries.

(iv) Possible simplification of shipping arrangements with reductions in costs and improved delivery times. Delivery times can be especially important where food is intended for food security-related emergency operations in deficit regions and areas.

(v) Improving the distribution system as a result of associated investments in infrastructure, system management and general operation in response to greater use.

(vi) Stimulating regional trade. The organisation of triangular operations confronts existing non-tariff barriers.
(vii) Providing a focus for international cooperation, bringing together food aid and financial and technical assistance.
(viii) Exchanges allow for additional commodity assistance when finance is not available for a local purchase and when imported food aid commodities are not appropriate for the intended use.

However, such arguments in favour of cash purchases and exchange arrangements should be counterbalanced by the hypothetical, and sometimes plausible, negative consequences of these operations, which include:

(i) Uncertain availability, possibly preventing the satisfactory completion of an operation even when a purchase or exchange contract has been formally signed. The low level of procurement through official channels, coupled with fluctuating production, can result in thin and volatile regional and national markets. The amount of food reaching the market in any one year can therefore potentially amplify the impact of the weather on the largely rainfed production of basic staples. This can result in the sudden cancellation of purchase contracts and inhibit multi-year arrangements.
(ii) Barriers to the systematisation of purchase and exchange operations. Fluctuating production levels also mean that purchases cannot always be made on a regular basis, but only intermittently, preventing the systematisation of purchase procedures.
(iii) Inadequate availability of information on local current and projected availability and on prices.
(iv) Potential quality problems.
(v) Increased delivery times. Operations may not be more timely and could involve unacceptable risks, particularly when the food is intended for emergency operations.
(vi) Increased costs. Triangular arrangements and exchanges may reduce transport costs but at the same time could involve additional (hidden) administrative costs in excess of savings.
(vii) Implicit support for agricultural and pricing policies. Financing additional purchases may sustain inappropriate agricultural policies.
(viii) Interference in domestic markets. Purchases may distort normal inter-regional food flows. Also, purchases in pre-harvest periods, when food stocks are at their lowest, may lead to price increases.

(ix) Interference in natural trade flows and food security practices. Stimulating regional trade could involve trade diversion against natural comparative advantage and divert food away from use in food security stocks.

(x) Misallocation of investment resources. Investment in strengthening certain aspects of the logistical system to facilitate cash purchases and exchanges may represent a mis-allocation of resources.

(xi) The promotion of developing countries' exports or local purchases with cash may involve a substitution of food aid for other forms of financial assistance.

(xii) Tying food aid to exchange and direct use of local food may be less appropriate than monetising it and allocating local currency to development activities. The latter may have a potentially higher return in productivity or the creation of human capital.

In view of the foregoing, assessment of the likely overall impact of a purchase or exchange must be based on a full and careful analysis of the specific economic and institutional context in which it is proposed rather than solely on general principles.

Recent Experience of Triangular Transactions and Local Purchases

Evaluations of recent experience with triangular and similar operations, particularly in SSA, have focused on aspects of cost effectiveness (in this context defined in terms of the import parity price for the importing region or economy), timeliness of delivery and appropriateness. Purchases from regular exporting developing countries were found to be similar to food aid from developed countries in terms of cost effectiveness and timeliness. But they were likely to result in the supply of a more appropriate commodity (for example white maize and sorghum, which could be purchased in Thailand). In contrast, purchases from non-regular exporting developing countries, which have been promoted as offering improvements in these areas, had a much more mixed record.

First, such operations can be, but are not always, cost-effective in terms of commodity and related transport costs in quite a wide range of cases. They are most likely to be cost-effective when actions involve trade between landlocked countries or neighbouring interior regions and when transport to and from ports is expensive and poor. Examples were bilateral actions organised by the WFP from Kenya to Southern Sudan and Southern Ethiopia. Many actions involving shipments of

Zimbabwean maize to neighbouring provinces of Mozambique were similarly found to be cost-effective.

However, as in some instances commodities acquired through purchase are transported over long distances to their final destination, lower transport costs *per se* are not an integral feature of triangular transactions; indeed transport costs could be higher than if the food were purchased from other sources. But the cost effectiveness of operations needs to be considered in terms of both commodity and non-commodity costs and whether they are at or below import parity prices for the final destination. On these criteria, purchases involving the shipment of pulses were found to be cost-effective because of the competitive export prices in source economies. The results were less convincing for some of the maize operations in 1986/87. And with higher international prices in 1988 different results might be obtained for that year.

Hay and others have been the only ones to examine the trend in the cost effectiveness of purchases over time. They found that between 1984 and 1987 there was an overall decline in the efficiency of triangular transactions in cereals relative to food aid from Europe. This must in part reflect the fact that the scope for cost-effective actions had been circumscribed by international grain prices. In 1986/87 these prices reflected competitive dumping/subsidisation by major exporters.[14] However, the world market of 1988 was much tighter as a result of the US drought and world and export prices were much closer to internal prices in some exporting countries. The extent of the production response in the main exporting countries to these high prices will heavily influence the short-term prospects for cost-effective triangular operations. Similarly, the outcome of the Uruguay Round negotiations, which resumed in April 1989, will have important consequences for the longer term.

Second, experience is more mixed with regard to the timeliness of delivery and assuredness of supply for triangular and similar type operations. A high proportion of WFP actions involving coarse grains in SSA went through quickly and compared favourably with typical times between requests for, or decisions to send, emergency shipments from major exporting countries and their arrival [*RDI, 1987*]. So, if food aid is seen as part of a trade-based strategy for food security, there would appear to be a case for making greater use of triangular operations. However, EC triangular transactions were not found to be more timely than bilateral food aid operations [*Hay et al., 1988*]. Various studies have highlighted a small number of purchase actions by different donors which were subject to lengthy delay, high administrative costs, eventual deferral and even cancellation. In some cases alternative normal food aid operations or an alternative triangular operation from Asia had to be organised [*Borton et al., 1988*]. These cases, and the high administrative costs of most triangular type operations, indicate policy issues that require further analysis.

Third, maize and other coarse grains provided under triangular type operations were, by and large, more appropriate than alternative commodities available from developed exporting countries. However, in some cases, as not infrequently with normal food aid, there have been serious problems with the quality and the phyto-sanitary condition of the grain made available by parastatals or private traders in, for example, Somalia and Sudan.

Fourth, triangular transactions have been successful in transferring resources to source countries, but these benefits have accrued mainly to governments rather than to the farmers, according to Hay and others, who considered the supply side arguments for cash purchase in greater detail.

Recent Experience of Commodity Exchanges

Trilateral exchanges as undertaken by Australia and the USA appear to perform similarly to triangular purchases. The crucial determinant of the cost effectiveness of local exchanges is the ratio at which commodities are exchanged. This can be based on a variety of prices or on volume or nutritional criteria. Obviously donors seek to negotiate exchange ratios which they consider to be fair to themselves as well as to the bodies with which they are dealing. However, there may be considerable time lapses between the negotiation and implementation of exchange ratios, during which time the relative prices of the commodities being exchanged may have changed dramatically. Thus local exchanges may not be more cost-effective than direct import and distribution of food aid although they do normally entail lower transport costs.

The timeliness of delivery of commodities supplied through exchange has been mixed. In some instances project authorities were able to draw down the exchanged commodity before delivery of the imported commodity, thus reducing delivery times. But, on the negative side, there may have been delays in negotiating the exchange agreement which would not have arisen with direct import and distribution of food aid commodities. In some cases, when exchanges were conducted through government parastatals, and when credits were held with them, it was possible to draw down commodities against these credits, so enabling a rapid response to sudden emergency needs. Credits were then replenished at a later date with imports. Thus, in some instances exchange arrangements already entered into facilitated a rapid response to a sudden need. However, in some countries problems were encountered with the availability of the exchanged commodity, sometimes even after the imported commodity for exchange had already been delivered. In such cases delays were possible in the delivery of food aid to the final recipients. However, exchanges certainly appear to have permitted the provision of food more in keeping with the tastes of final food aid recipients.

POLICY AND RESEARCH IMPLICATIONS

Lessons from Recent Experience

The tendency for commodities acquired through triangular and local purchases, and particularly those from non-regular exporting developing countries, to be used for emergencies rather than in development projects has already been mentioned. This pattern is indicative of the thinness of markets within SSA and the quite rapid changes in the supply and demand situation that can be expected. Surpluses can often not be anticipated and so not incorporated into long-term allocations to multi-year development projects, but instead have to be used opportunely, matching surpluses with deficits as and when they arise. This *ad hoc* approach inevitably entails problems not encountered in using the regular surplus sources. Just when there is pressure to carry through the transaction as quickly as possible because of the emergency need, unfamiliar purchasing and export formalities are encountered which will entail almost inevitable delays. These, combined with sudden shifts in commodity availability, mean that purchase contracts may be only partially honoured.

In fact, the typical emergency response time for the supply of food from developed countries is six months (excluding small, highest priority, shipments). However, in practice many actions take longer, especially to landlocked destinations. Triangular operations have not succeeded in reducing this time significantly. The implication is that national distribution systems must maintain some combination of working/food security stocks to provide a margin of assurance in relation to a realistic assessment of such lead times.

The predominant use of commodities purchased in developing countries for emergencies rather than for development projects reflects the fact that emergency and disaster relief funds are often more

TABLE 6

TOTAL FOOD AID AND DEVELOPING COUNTRY CASH PURCHASES AND TRILATERAL EXCHANGES OF CEREALS (AS REPORTED TO THE FAO) ARRANGED BY DONOR IN DECREASING IMPORTANCE OF INVOLVEMENT IN PURCHASES/EXCHANGES, 1983/84 TO 1987/88 (IN GRAIN EQUIVALENT THOUSAND TONNES)

	TOTAL FOOD AID	DEVELOPING COUNTRY PURCHASES/TRILATERAL EXCHANGES	PURCHASES/EXCHANGES AS % OF TOTAL FOOD AID
JAPAN	2265.8	1269.2	56.0%
EC COMMUNITY ACTION	5567.4	627.8	11.3%
NETHERLANDS	721.1	464.7	64.4%
GERMANY, FR	1159.0	351.6	30.3%
UK	730.7	195.8	26.8%
SWITZERLAND	223.8	175.0	78.2%
NORWAY	193.5	163.4	84.5%
AUSTRALIA	1967.5	126.4	6.4%
DENMARK	98.8	99.8	101.0%
CANADA	5277.6	95.4	1.8%
AUSTRIA	96.8	60.4	62.4%
USA	35673.8	60.4	.2%
BELGIUM	214.7	41.2	19.2%
ITALY	605.4	33.8	5.6%
OPEC Special Fund	26.2	24.8	94.8%
OTHERS (a)	4040.8	155.101	3.8%
WFP PURCHASES	191.1	189.9	99.4%
TOTAL	59054.0	4134.809	7.0%

Source: Unpublished FAO data
(a) Other countries and Community Funds, ICRC, NGOs, UNHCR, UNKAM and WFP (insurance claims).

flexible than development funds. They can be used more easily in an *ad hoc* response to the sudden availability of commodities in a developing country. It is also more practical from the donor point of view to make developing country purchases to meet sudden emergencies than to enter into longer-term, multi-year agreements. But the producer has no guaranteed outlet for agricultural surpluses in future years.

The partial exception is the WFP. In countries where supply is more certain WFP agreements on food aid projects with recipient governments sometimes explicitly include arrangements for local purchase or local exchange as the means of supplying certain commodities. In agreements with countries where supply is less certain, exchanges have been specified as the source of certain commodities, with contingency plans for the same commodities or acceptable alternatives to be imported directly should they become unavailable locally.

A further question is whether the performance of triangular transactions and local purchases improves over time as they become routine operations. Since their ultimate operational credibility depends in part on this issue, it would be desirable to monitor experience of this aspect since 1986 for selected important source countries.

Triangular type operations are attractive to parastatal organisations and governments as an alternative to meeting the costs of inter-year storage. Where parastatals are under pressure to improve their budgetary position as part of an adjustment programme there is a powerful incentive to engage in short-run disposal measures. The willingness of donors to meet administrative and transport costs shifts the relative costs and advantages of storage versus disposal; and some governments, such as those in Malawi and, possibly, Niger may have been tempted into such apparently attractive short-run practises. For example, under triangular agreements made in 1985 Malawi continued to export maize until 1987. But it is now importing maize itself after the unexpected influx of Mozambiquan refugees and in the face of domestic problems. Were those aid-financed exports in 1986, at export prices that could compete in, for example, the Botswana market, sufficient compensation for commercial import/food aid at Malawi's own high import parity prices in 1987? It is hard to avoid the conclusion that donor responses have been reactive and not based on a fuller assessment of the food security implications for the exporting economies.

The considerable non-tariff barriers to legal trade in basic staples within countries and across borders have been highlighted by attempts to organise triangular operations [*Club de Sahel, 1988; RDI, 1987*]. Non-tariff barriers such as, for example, obstacles to the private trade in grain from region to region within a country, in part reflect national food security concerns. And cross-border trade in basic staples may be more circumscribed by such food security-related barriers than traditional trade in export commodities. The barriers in exporting countries are duplicated by the regulations of importing countries, particularly as these affect commodities crossing land borders. In some cases the extent of regulations imposed by the importing country suggests that countries may not be serious about developing trade when this implies a reliance on a neighbouring economy for basic staples.

In part, barriers reflect justifiable concern over phyto-sanitary problems, as at present, for example, in relation to greater grain borer infested maize that might otherwise be exported from Tanzania. However, problems encountered in moving maize overland from Kenya to Ethiopia or from Mali to Senegal in 1986/87 appear to have reflected unjustifiable concern about phyto-sanitary conditions and in reality to have been part of an administrative reluctance to allow such movements. The barriers actually reflect the generally complex and weak trade regulatory administration in potential exporting countries. If legalised trade in basic staples is to grow in importance between SSA countries, substantial measures are required to reduce unnecessary non-tariff barriers. For example, complementary measures are required to ensure the quality and phyto-sanitary condition of grain stocks.

Meanwhile, the increasing role of local purchases creates an ambiguity in food aid statistics, since a significant part of assistance to some African countries no longer involves food imports. As a consequence, food aid can no longer be directly estimated as a proportion of total recorded imports in any meaningful way for, for example, a number of Sahelian countries and Sudan.

Implications for Aid Donor Support for Food Security

The purchase of food aid commodities from within SSA has been justified strongly in terms of the developmental benefits to the source economy and the economic gains from stimulating regional trade. Yet that aspect of operations has received the least attention in the economic analysis of actual operational experiences to date (except by Hay and others). Donors have been largely reactive, responding to the short-term opportunities for food aid actions provided by the changing supply situation in source countries and regions rather than having longer term regional trade strategies in mind. A large part of these commodities appears to have been used in emergency operations, notably in Mozambique, Sudan, and Ethiopia, but also elsewhere. Such operations are by definition supposed to be *ad hoc* and reactive, although actual practice is somewhat different, as, for example, in the on-going emergency nature of the situation in Mozambique and in Somalia, where there are long-term refugee/displaced populations. But, if commodity acquisition in developing countries is to have a continuing and important role in food aid, it would seem imperative for donors, in co-operation with potential source economies, to develop a coherent medium-term framework for planning and organising them.

To address this issue, the RDI [*1987*] study made a set of specific suggestions for improving both the planning and management of triangular and local purchase arrangements that would concern both donors and beneficiary countries. These included improved trade information, removal of non-tariff barriers, etc. Hay and others made similar proposals.

However, there are problematic implications if donors become involved in the planned promotion of regional trade in food. The ongoing review of the operational elements of food security in the SADCC region, and the deliberations over food trade policy for the Sahel region at the Mindelo Conference, have highlighted some of these issues, such as how to distribute the benefits and costs of the trade creation and diversion effect of a putative protected Sahelian regional cereals market [*Club du Sahel, 1987*].

The World Bank and other donors are involved in policy analysis and in supporting the strengthening and reform of grain-marketing operations in a number of countries. Recent experience with triangular operations has highlighted the complex and considerable non-tariff barriers to trade that have ambiguous and often negative food security implications. This is a potentially fruitful area for policy reform. There are capital investment and training implications if trade is to be promoted in this way. A theme of the early 1980s in donor policy discussions was that food aid should be programmed to support, or be consistent with, financial aid. The implication here is that other aid resources may be required to complement the developmental use of donor food purchases, especially in SSA.

There is a danger of donor policies amplifying the basic cereals cycle in relation to thin markets. As a result of their slowness to programme and implement food aid actions, and of barriers to trade, some of their interventions may even become pro-cyclical. Policy reform should presumably address stockholding decisions in terms of a realistic assessment of costs versus risks.

The priority accorded to national food security where storage of and trade in cereal staples are involved makes it unlikely that countries will go far in liberalising trade in food commodities with neighbouring countries. Are there ways in which politically attractive donor interventions can be better integrated into a medium-term programme for strengthening food security and marketing reforms directed to agricultural development?

Implications for Further Research

The large-scale acquisition of commodities in developing countries by food aid donors is such a recent phenomenon that an analysis of its impact and implications must be regarded as provisional. The following issues should receive particular attention in future research.

First, how does the provisional conclusion that such operations were broadly cost-effective hold up through the period of volatility in international cereal prices of recent years? The issue of price determination in administered trade is difficult and politically sensitive, and needs further consideration.

Second, the intended developmental benefits of donor commodity acquisition have yet to be systematically reviewed, except provisionally for Zimbabwe and SADCC countries. There are some difficult analytic and empirical problems to be addressed.

The analysis of the consequences of donor intervention in a national or regional food system where there are official and parallel markets and complex non-tariff barriers to trade is a challenging issue.

NOTES

1. Some NGOs describe commodities acquired in Sudan for cross-border operations in Tigre and Eritrea as local purchases when these are a form of triangular transaction.
2. For example, the United States and Australia have made normal food aid shipments of wheat to Zimbabwe. This has been exchanged for white maize which has then been shipped at the donor's expense as food aid to another developing country.
3. For example, WFP/CFA [1989] includes Malawi and Zimbabwe amongst regular cereal exporters while the authors classify them as non-regular.
4. The Grain Marketing Board is part of a government parastatal with the responsibility for ensuring sufficient domestic food crop supplies.
5. All quantities are reported in cereal equivalent tonnes.
6. In the Clay and Benson study [1989] problems were encountered in disaggregating the total volume of WFP exchanges by year. Although the figures do not suggest an upward trend in the volume of exchanges as reported by the year of the first shipment, it is possible that there was an increase in the actual volume of commodities exchanged between 1985 and 1988.

7. This section draws on the data sources identified above (Evaluations). Unless otherwise stated, all discussions refer to FAO data for the years 1983/84 to 1987/88.
8. However, between 1982/83 and 1983/84 the volume of cereals cash purchases and trilateral exchanges fell by about a third.
9. However, Hay and others [*1988*] found that the volume of triangular transactions by the EC reached a peak in 1986 but declined in 1987 despite the Community's increasing support for them. This was possibly due to management capacity constraints amongst other factors.
10. In 1987/88 the source of 14 per cent of total purchases and trilateral exchanges was reported as unknown. Thus these figures may be even higher than stated here.
11. Figures on the commodity composition of cereals food aid refer to the years 1983/84 to 1986/87 only, since data for 1987/88 are not available.
12. This result is confirmed by Hay and others [*1988*] who found that, while maize dominated EC triangular transactions in eastern and southern Africa, beans and pulses were important in Latin America, and in West Africa and the Sahel a mixture of beans, sorghum, millet and maize. However, triangular transactions by other organisations using EC funds did not reflect this pattern.
13. In the past, domestic production and stocks of rice had a significant effect on the level and composition of Japanese food aid. A policy of restricting rice production was then introduced to reduce stocks while still maintaining self-sufficiency. When the Japanese food aid programme could no longer draw heavily on domestic rice stocks it embarked on its policy of food purchases from developing countries.
14. In such years, if calculations were to be based on international prices that more closely reflect production costs in major exporting countries, the economic case for more integration of grain markets within SSA would probably be greater than suggested by cost-effectiveness analysis of recent triangular operations.

REFERENCES

Borton, J., et al., 'ODA Emergency Aid to Africa 1983–86', *Evaluation Report* No.EV 425, August 1988, London: Overseas Development Administration.
Clay, E.J. and C. Benson, 1988, *Evaluation of EC Food Aid in Tanzania*, London: Relief and Development Institute.
Clay, E.J. and C. Benson, 1989, *Study of Commodity Exchanges in WFP and Other Food Aid Operations*, London: Relief and Development Institute.
Food and Agriculture Organisation, 1987, *Food Aid in Figures*, No.5, Rome: FAO.
Hay, R., B. Martens and S. Hunt, 1988, 'European Community Triangular Food Aid: An Evaluation', Draft Synthesis Report, Oxford: Food Studies Group. Dec.
IDS/CEAS, 1981, *The European Community's Cost of Food Aid Study*, Vol.1: 'Main Report' (E.J. Clay and M. Mitchell); Vol.2: 'Country Case Studies: Bangladesh, Lesotho, Mauritania, Sri Lanka', University of Sussex: Institute of Development Studies/University of London: Centre for European Agricultural Studies.
Jost, S., 1985, 'Etude d'évaluation prospective sur les opérations triangulaires d'achat de l'aide alimentaires sur les marchés locaux en Afrique de l'Ouest', paper presented at a meeting of the Réunion du Réseau pour la Prévention des Crises Alimentaires dans le Sahel, Paris, 23–24 Oct.
Jost, S. and J.-J. Gabes, 1988, 'Food Aid to the Sahel: the Situation in 1987/88'. Club du Sahel-OECD, Nov.
Relief and Development Institute, 1987, 'A Study of Triangular Transactions and Local Purchases in Food Aid'. Paper commissioned for WFP, *Occasional Paper* No.11, Rome: WFP, July.
Ronco Consulting Corporation, 1988, 'Trilateral Food Aid Transactions: US Government Experience in the 1980's'. Contract PDC-1096-I-03-4164-00, Washington, DC: USAID.

Shaw, D.J. 1983, 'Triangular Transactions in Food Aid: Concept and Practice – the example of the Zimbabwe operations', *IDS Bulletin*, Vol.14, No.2: 29–31, University of Sussex.

World Food Programme (WFP), 1974, 'Resolution XVIII of the 1974 World Food Conference – An Improved Policy for Food Aid', IGC/CFA, Rome.

World Food Programme, 1979, 'Guidelines and Criteria for Food Aid', WFP/CFA, 7/21, Annex 1V: 47, Rome.

World Food Programme, 1988a, 'Food Aid Policies and Programmes', Rome, May.

World Food Programme, 1988b, 'Annual Report of the Executive Director of the World Food Programme for 1987', WFP/CFA: 25/P/4: 31–35, Rome.

7

Food Aid and Structural Adjustment in Sub-Saharan Africa

HANS SINGER

A CONVERGING CONSENSUS

Over the last decade of this century the role of food aid in structural and sectoral adjustment support (SAL) for sub-Saharan Africa will be a matter of greatly increased importance. This is the result of a convergence of three factors:

1. A consensus that the problems of sub-Saharan Africa (SSA) are of such an exceptional severity and unique nature as to require and justify new approaches and to impose an obligation to leave no potentially useful resources unused. This consensus includes agreement on special treatment for SSA in respect of debt reduction, and special facilities have been readily accepted;
2. A consensus that adjustment and sectoral support in the case of SSA must be more 'growth-oriented', must have a more 'human face' and require both a longer-term horizon and increased resources;
3. A consensus that there is both need, and potential absorptive as well as financing capacity, for additional food aid, subject to an assurance that food aid helps to provide incentives for increased food production and physical and human investment in recipient countries.

Standing as it does at the crossroads of such increasingly accepted policy orientations, the subject clearly deserves new thinking preparatory to constructive action.

QUANTITATIVE DIMENSIONS

As a starting point for visualising the quantitative scope of such constructive action over the next decade, we take the International Food Policy Research Institute (IFPRI) estimate of cereal food aid needs for low-income SSA countries in 1990 of 11.36 million metric tonnes [*IFPRI, 1988*].[1] Compared with a present annual flow of four to five million tonnes (varying with the acuteness of emergencies and international response), this would require an expansion by some seven tonnes per annum, increasing the current total flow of food aid from ten to 17 million tonnes.[2] There is no doubt that such additional supply capacity exists (some of the extra seven million tonnes could come from within SSA by way of triangular transactions). Given the proposed link with SAL, much of this additional seven million tonnes would take the place of present commercial food imports, freeing foreign exchange for developmental imports (targeted towards stimulating domestic food production) and, it is hoped, reducing food aid needs towards the end of the decade. The problem is not one of supply capacity but of aid capacity: even a doubling of food aid would leave total aid flows at 0.39 per cent of donor GNP (instead of the present 0.35 per cent), still a long way below the 0.7 per cent UN target; it would do little more than restore food aid to the level of 25 years ago and bring it to the one-sixth of total aid flows proposed at that time by expert analysis [*Chakravarty and Rosenstein-Rodan, 1965*]. It would also do no more than maintain the present share of food aid (about 20 per cent) in the rising cereal import requirements projected on present trends for SSA in the 1990s. It would represent four per cent of present stocks, or six per cent of world cereal trade, or 0.02 per cent of donors' GNP.

However, the real limit is absorptive capacity, in the positive sense of making food aid an efficient instrument for bringing SSA back towards renewed growth and development. It is here that its integration with the present SAL efforts becomes vital. In this chapter we shall explore ways of doing this, and it will emerge that the possibilities exist, to the mutual benefit of SAL and food aid, of making both into better instruments of support for SSA.

Any action would, of course, start on a smaller, more experimental, scale than the estimated ultimate limit of seven million tonnes annually. It would also be largely on a country-by-country and trial basis, without necessarily any quantitative target. A more modest proposal for immediate action involves the creation of a new international facility of $1 billion a year; the 50 per cent of this proposed as food aid (and this not all for SSA) would represent at present prices about 1.5 million tonnes of cereals, one-fifth of the potential scope estimated above [*Reutlinger, 1988*]. This proposal is rightly described as 'obviously only a beginning, though a respectable one' [*Reutlinger,*

1988: 64]. At this stage, clarity about objectives and ways to integrate food aid and financial aid, and the achievement of efficient income transfer, seem more important than discussion of quantities.

A MENU OF POTENTIALITIES

Food aid can contribute to structural adjustment in nine different ways.[3] By:

1. contributing to overall available resources, to cushion the austerity of the adjustment process;
2. setting free foreign exchange, helping with the stabilisation of the balance of payments as a necessary preliminary to resumed economic growth;
3. sustaining incomes and employment of the poor, essential during the difficult transition period from retrenchment to resumed growth;
4. protecting vulnerable groups from the harsh short-run impact of stabilisation and adjustment policies;
5. maintaining essential economic and social services during the transition period;
6. reducing budget deficits and inflation – an important element in stabilisation and adjustment;
7. helping to provide domestic food stability and food security in SSA – an essential political precondition for making adjustment acceptable and viable;
8. supporting essential reform measures and sectoral programmes often related to food production;
9. helping, through counterpart funds, to provide local finance for projects or programmes within a structural adjustment context.

With such a multitude of possible objectives and effects, it is important to be clear about intentions, and to choose the forms of food aid and financial aid, and other methods of combination, in the light of the main objectives. It will be useful to look briefly at these objectives in turn.

Increasing Overall Resources – Cushioning Austerity

It is increasingly recognised that the SSA countries must be given a greater incentive as a *quid pro quo* for the dangers, risks and uncertainties that are involved in the short-term (stabilisation) part of the adjustment process. Experience has shown that many programmes fail at this first hurdle. Most SSA countries are so poor, and in some of them the political, social and economic structure is so fragile, that the promise of better things to come may be heavily discounted. 'In the long run we are all dead' and the run may be shorter in SSA than elsewhere. So any increase in the *quid pro quo* would be welcome, particularly if front-loaded to come during that difficult initial period. To be front-loaded, the additional food aid would normally have to be on a quick-disbursing programme basis; or else given to countries where an infrastructure and absorptive capacity for additional food aid already exists.

As an increase in overall resources, the 70 per cent expansion of total food aid from ten to 17 million tonnes previously mentioned would represent $2 billion per annum; the same order of magnitude as some of the other provisions and facilities which have been suggested for sub-Saharan Africa in the way of additional structural adjustment funds, additional aid, etc. But the more realistic suggestion of 500 million per annum would clearly be a subsidiary factor taken as a simple resource flow. To take account of the minority of food-exporting SSA countries, this flow would have to be mixed with fungible financial resources and/or partly on the basis of triangular transactions; although theoretically, food-exporting countries could also benefit from increased food aid made possible by switching resources into increased exports, either by way of increased investment or by switching production out of food-aid commodities. (If they are genuinely and viably food-exporting, this need not endanger their own food security.) From this angle, food aid and related financial aid should be targeted to countries in greatest danger from initial austerity or those where scarcity of overall resources is the main bottleneck preventing growth, that is, countries with unutilised production capacity, unemployment or unused land, or where inputs and investment do not require foreign exchange. This particular purpose is frustrated if current food aid is merely switched into financial aid associated with SAL (although it could increase the efficiency of the resource transfer).

The Usual Marketing Requirements (UMR) guidelines are designed to ensure, where this is intended, that food aid is additional to commercial imports; but they work only patchily and may be in need of clarification in the case of food aid given as part of a structural adjustment package.

Freeing Foreign Exchange – Balance of Payments Support

This will generally be the more important contribution of additional food aid. Practically all SSA countries are in a foreign exchange constraint more binding than the savings/investment constraint. To satisfy this purpose, food aid would have to displace commercial imports and be in general on a programme basis. Food-exporting countries would not benefit directly. The associated financial aid and conditionality would be designed to make certain that good developmental use is made of the foreign exchange set free, and in particular that it helps with sectoral programmes and projects which are part of the overall adjustment process. If the freed foreign exchange is added to reserves or used for debt repayment, it will not help directly to make adjustment easier or more growth-oriented; but it may help indirectly by increasing creditworthiness and making it easier to restore capital flows or obtain import credit. But this can hardly be a direct objective.

To the extent that it replaces commercial imports, food aid is fully equivalent in value to financial aid for the recipients, but for the donors the cost of transporting and delivering the food aid (itself requiring associated financial aid) has to be treated as an extra cost factor. In so far as the displaced commercial imports come from other SSA (or other developing) countries, there would have to be triangular transactions or additional financial aid for the food-exporting SSA countries involved – not just for reasons of equity, as in the preceding case, but in direct compensation for actual loss suffered. For the donors, the aid given would be at the cost of reduced commercial exports, but this would be compensated for by other exports paid for with the foreign exchange set free, for example, agricultural inputs or inputs needed to utilise capacity made idle by lack of foreign exchange – a widespread fact of life in SSA today.

Sustaining Income and Employment of the Poor

Two related criticisms of present SAL approaches are (a) that they are unduly restrictive in the initial period, imposing sacrifices well before any benefits can be expected; and (b) that these sacrifices tend to fall most severely on poorer sections, increasing inequality of income distribution. Increasingly, these worries are accepted as matters of legitimate concern. Food aid is particularly well suited to respond to such concerns. It is aid with built-in targeting towards the poor, who spend a larger part of their incomes on food; as the classical 'wage good' or 'subsistence fund' it also has a built-in tendency to favour labour-intensive activities; it is well suited to relieve urban poverty directly – it is the urban poor who are most affected by the need to give higher priority to agriculture and exports and to fulfil other SAL conditions; through monetisation and use of counterpart funds food aid can finance rural public works projects, which create demand for local food, and help to provide the infrastructure, such as improved rural roads, needed to support higher prices as an incentive to food producers. In many developing countries, including SSA countries, there is experience and an established administrative structure for such food-for-work or food-aid financed cash-for-work projects.

With such multiple advantages, and its ability to address what are increasingly recognised as problems in SAL procedures, this potentially fruitful combination of food aid and SAL is crying out to be more systematically exploited. In SSA, the threat to incomes and employment during the transitional stabilisation period is super-imposed upon a long-run decline, in some countries dating back as much as twenty years. Moreover, there is a danger that a decline in incomes and employment in the initial period may become cumulative and create a downward spiral, frustrating the objective of structural adjustment 'to lay the foundations for subsequent viable growth'. Timely support for employment and incomes can prevent such a vicious circle; but for this purpose it would be essential for food aid to be included at an early stage, that is, as part of the stabilisation as well as the adjustment package, rather than added later as a compensatory afterthought. In this case also, prevention is better than cure. But the compensatory function of food aid also has a justification of its own, discussed in the next section.

Protecting Vulnerable Groups

Apart from the recognised need to protect the poor in general during periods of adjustment [*World Bank, 1987*], there is a further need to protect specially vulnerable groups among the poor. The best-known exposition of this case in relation to children is the plea by the United Nations Children's Fund (UNICEF) for 'Adjustment with a Human Face' [*Cornia, Jolly and Stewart, 1987*]. The destruction of human capital in SSA as a result of the economic and debt crises, often reinforced by the initial austerity of the adjustment process, is not only unacceptable on humanitarian grounds, it is incompatible with the objective of adjustment, which is to lay the foundations for viable growth. In addition to sound policies, this objective will certainly require a well-fed, healthy, educated and skilled future labour force. It is now known that malnutrition and the associated ill-health of young children can have a permanent effect on their intelligence and future productivity. Between birth and the age of five a child does most of its learning and develops practically all its brain potential. While it is one of the declared objectives of adjustment to improve the efficiency of social expenditure, such as that on education, malnutrition among children reduces school attendance, increases drop-out rates and makes them less able to benefit from education. Food-aided child/mother health clinics, provision of school meals, nutrition education, and direct nutritional supplements can have high direct benefit/cost ratios (as the World Bank's own research has established) and even greater indirect, not easily measurable, benefits. Here food aid has a well-established tradition and infrastructure with a strong multilateral basis in WFP and UNICEF. Its association with the adjustment process will not only give adjustment 'a human face' and make it politically more acceptable, it will also make it more efficient and logically consistent. Destruction of human capital as part of the adjustment process defeats its own purposes.

Protection of vulnerable groups, with the help of food aid, is also a traditional and suitable field for non-governmental organisations (NGOs) who may be better equipped for close targeting. The involvement of NGOs is a declared objective of the providers of both adjustment assistance and food aid.

Maintaining Essential Services

SAL programmes often emphasise the increased cost efficiency of public services, usually in the context of reduced overall expenditure and increased taxation, often involving direct user fees. Governments may interpret the mandate of increased cost efficiency as the need to cut down services in rural areas, especially remote areas where they may be most costly to deliver, and concentrate on economic 'directly productive' services. They may do this to the detriment of essential social provision (endangering human capital), by imposing indiscriminate user fees (making access to public services less equal), and by cutting down on essential current and maintenance expenditure (resulting in clinics without drugs, schools without blackboards, agricultural marketing boards without spare parts for broken-down lorries or without fuel to run them, etc.). Short-run cost efficiency in the accounting sense is not necessarily true long-run developmental cost efficiency.

Such interpretations of cost efficiency may not be in line with the preferences of those providing adjustment support, but the alternative would be increased interference with the right of sovereign governments to have their own priorities. The declared principle of adjustment is to limit conditionality to macro-economic quantities but leave distribution within these quantities to the judgment of the government. Food aid provides a means of preserving this principle while giving governments both incentives and the means to avoid sacrificing true long-term cost efficiency.

Reducing Budget Deficits and Inflation

Programme food aid represents an important source of government revenue in some SSA countries. At the same time, by drawing money out of the private sector it reduces inflationary pressure, which is an objective of many SAL agreements. Higher food prices are one of the most potent sources of inflationary spirals which frustrate many SAL policies such as devaluation and other measures designed to improve the balance of payments and foster an outward orientation. The revenue from food aid sales can also be used to promote local food production by paying higher prices in a segmented food market, or by spending on rural infrastructure and agricultural research and extension, by subsidising inputs or by increasing the supply of incentive goods in rural areas. All this can make the general SAL policy of higher prices for farmers more effective. Food aid fills a budgetary gap created by selective food subsidies, and prevents the diversion of government revenue from essentials for production growth.

The contrary fear, that the lowering of food prices acts as a disincentive for local farmers, should be ruled out if the food aid is given in the framework of a SAL programme which shifts incentives in favour of the rural sector. In any case, both historical experience, from the Marshall Plan to India, as well as economic analysis, contradict the view that such a disincentive effect necessarily occurs. An analysis of 16 developing countries which achieved particularly high growth rates in food production between 1961 and 1976 shows that they received 80 per cent more food aid per capita than the average food-aid recipient country [*Bachman and Paulino, 1979*]. Much of the fear of a disincentive effect stems from a mistargeting of criticism at food aid instead of at the surplus-creating agriculture and protectionist agricultural policies of the industrial countries, and neglect of the agricultural sector in developing countries. (Structural adjustment ought to deal with the first as well as the second factor.) Another source of fear of disincentive effects is simplistic market analysis which ignores the dual market structure in nearly all developing countries.

Providing Domestic Food Security

Food riots resulting from food insecurity are one of the main political obstacles to the acceptance, and even more the implementation, of structural adjustment programmes. Food aid flexibly and fungibly mixed with financial aid and help with the establishment of stabilisation stocks can overcome this problem. The food-aided rice price stabilisation stock set up with WFP co-operation as part of a World Bank-supported programme of agricultural reforms in Madagascar provides an example which should be capable of replication in other SSA countries [*Nicholas, 1988*]. Such arrangements require multi-annual planning of both the food aid and the financial aid components of the support package, with fungibility between the two in the sense that, at a time of sufficient stocks, aid assumes more financial forms, and *vice versa*. Food security will often require the maintenance of selective and targeted food subsidies; the gradual reduction in and phasing out of food subsidies is greatly preferable to abrupt abolition; and there is already African precedent for using food aid in a World Bank sectoral adjustment loan package to fund at least half of a targeted food subsidy scheme.[4] The EC pioneered the fungible substitution between financial and food aid when in 1984 it introduced its 'substitution accounts', replacing food aid with financial aid or technical assistance when the supply of food is not necessary or desirable [*Franco, 1988*]. Another contribution to food stability could be by flexible commodity aid that can be switched between food, fertiliser, seeds, pesticides, livestock feed, etc. The use of the International Monetary Fund's Compensatory Financing Facility could also be a flexible alternative to food aid; it now provides for a temporary excess in the cost of cereal imports but its use has been restricted by tighter conditionality.

Supporting Essential Reforms

SALs typically include measures such as liberalising food trade (abolition or reform of parastatal marketing boards), reduction of food subsidies, land reform, resettlement, and civil service reform (often involving reduction in numbers employed). All these measures, which involve some initial disruption and costs before the benefits show, reflect the general nature of the stabilisation/adjustment process, which is that the bad news (retrenchment) comes first while the good news (better-founded growth) comes later. Food aid can help to resolve the stresses resulting from this unfortunate sequence in the case of specific reforms just mentioned. The use of food aid in the initial stages of, for example, land reform, resettlement and food subsidies is well-established, and many people in the WFP and elsewhere have much experience of such schemes in SSA and other areas. In other cases, such as civil service reform or the liberalisation of food trade, the use of food aid would require new approaches – first probably on a pilot or experimental basis. This application of food aid may apply specifically to the sectoral rather than overall adjustment programmes (which, in fact, represent the bulk of SAL). For example, food aid seems well designed to maintain government workers laid off as a result of public sector reform, and to help in their retraining and, where it applies, their rural resettlement.

Using Counterpart Funds

Counterpart funds from the sale of food aid provide an opportunity for that continued dialogue with the recipient government which is so essential to a harmonious adjustment process. They can be used to help finance rural public works projects, thereby increasing the income transfer efficiency of food aid, and to augment demand for local food and offset any 'urban bias' or disincentive effect. In other words, counterpart funds can be considered a normal form of monetised food aid. Conversely, counterpart funds arising from financial aid can provide the local finance for food-aided projects, especially if this also serves as a catalyst for the (often very modest) foreign exchange required. The application of counterpart funds would also be a natural basis for continued dialogue, not only with the government, but among all the sources of the financial and food aid that gives rise to such funds. The best arrangement would clearly be multilateral coordination through the World Bank (directly or through the Consortia and Consultative Groups) and the WFP (directly or through the UNDP Round Tables). Such multilaterally coordinated administration, already shaping up in a number of SSA countries, could make an important contribution to successful adjustment, helping the region back to the path of development.

CONCLUDING REMARKS

This chapter is limited to the use of extended developmental food aid in adjustment. Unhappily, much of the food aid to SSA in the next few years will still have to be labelled 'emergency'. This is not to say that a broadened and more effective system of emergency food aid cannot also make an important, indeed indispensable, contribution to adjustment (which is itself a form of rehabilitation from Africa's 'chronic emergency'). But even with this limitation it has become apparent that there are many promising openings. No doubt this realisation will bring to light problems and difficulties but the potential benefits seem too great to be ignored. Multilateral and multilaterally coordinated food aid in the service of growth-oriented adjustment would also remove food aid from its unwanted link with surpluses and agricultural export wars.

NOTES

1. The figure appears in the section on 'Food Aid Trends and Projections' by Hannan Ezekiel. Estimates of food aid needs, prepared by the US Department of Agriculture and by FAO broadly agree that even on a *status quo* basis food aid needs exceed current availabilities by some 50–80 per cent. On a basis of nutritional needs the required expansion would be vastly larger but neither the authors of these projections nor of this paper assume that the task of eliminating hunger should be solely assigned to food aid.
2. This figure also is close to results obtained by USDA and FAO.
3. This list is based on a similar enumeration of the functions of food aid in Shaw and Singer [*1988*] and Sukin [*1988*].
4. The reference is to Morocco, which is not part of sub-Saharan Africa.

REFERENCES

Bachman, K. and L. Paulino, 1979, 'Rapid Food Production Growth in Developing Countries', *Research Report* No.11. Washington, DC: International Food Policy Research Institute.
Chakravarty, S. and P.N. Rosenstein-Rodan, 1965, *The Linking of Food Aid with Other Aid*, Rome: Food and Agriculture Organisation, p.22.
Cornia, G.A., R. Jolly and F. Stewart (eds.), 1987, *Adjustment with a Human Face*, Oxford: Clarendon Press.
Franco, M., 1988, 'Food Security and Adjustement: The EC Contribution', *Food Policy*, Vol.13, No.1.
International Food Policy Research Institute, 1988, *Third World Food Markets*, Policy Brief No.2, Washington, DC.
Nicholas, P., 1988, 'Adjustment and the Poor: the Role of the World Bank', *Food Policy*, Vol.13, No.1.
Reutlinger, S., 1988, 'Efficient Alleviation of Poverty and Hunger – A New International Assistance Facility', *Food Policy*, Vol.13, No.1.
Shaw, D.J. and H.W. Singer, 1988, 'Food Policy, Food Aid and Economic Adjustment', *Food Policy*, Vol.13, No.1.
Sukin, H.E., 1988, 'US Food Aid for Counries Implementing Structural Adjustment', *Food Policy*, Vol.13, No.1.
World Bank, 1987, 'Protecting the Poor during Periods of Adjustment', paper for a meeting of the Development Committee, Washington, DC, April.

8

An Introduction to the Sources of Data for Food Aid Analysis with Special Reference to Sub-Saharan Africa

STÉPHANE JOST

INTRODUCTION

This chapter aims to inform readers who want to undertake their own analysis of food aid, or to explore further some of the analyses included in this volume, about the sources of statistical data on global food aid and donor organisations. The data also enable the researcher to build up a picture of food aid for individual recipient countries, each of which has its own distinctive statistical system. Lastly, the reader should note that some of the sources of information are restricted. This means that they are only circulated to aid agency officials and some NGOs. Nevertheless, researchers have on occasion had access to some of these data.

Many agencies collect data and provide statistics on food aid. But often their aims are different and the mode of presentation varies from one source to another, making comparisons between series very difficult. The main potential sources of statistical data on food aid and their chief characteristics are described here. Some of the difficulties encountered in using and comparing the series are noted.

Such an exercise points to some possible improvements and the need for greater co-ordination between the agencies in collecting and distributing data. International guidelines to facilitate the comparison of statistical data, and the follow-up of each food aid allocation, are proposed.

SOURCES OF INFORMATION ON FOOD AID

Food and Agriculture Organisation (FAO)

FAO provides a number of important data series. Its Commodity and Trade Division's Food Security Group is responsible for publishing *Food Aid in Figures*, which has become, since it appeared in 1983, the single most widely available source of statistical information on food aid collated from many sources. Initially an annual publication, it now appears bi-annually in January and July and includes data previously published quarterly in the FAO's *Food Aid Bulletin*.

The geographic coverage is global and the data go back to 1970/1971. For each beneficiary it gives shipments of food aid for the last 12 years in cereals, by type of cereal, as well as in non-cereal commodities. For the most recent year, food aid flows for each commodity are also detailed according to the bilateral donor who funds (even multilateral) food aid, but is not necessarily the source of commodities. The food aid provided through international agencies (World Food Programme, United Nations High Commissioner for Refugees, United Nations Relief and Works Agency, United Nations Children's Fund, International Committee of the Red Cross, League of Red Cross and Red Crescent Societies) is summarised. Information on donor food aid budgets, and transactions reported (as discussed below) to the Committee on Surplus Disposal and the International Wheat Council under the Food Aid Convention, are also reported. The Food Security Group has also tracked triangular transactions and local purchase operations since 1983/84 and these are reported in *Food Aid in Figures* and GIEWS publications (see below).[1]

The Committee on Surplus Disposal (CSD), a Washington-based bureau of FAO, registers the food aid allocations of donor countries for comparison with the usual level of recipients' commercial imports. It then determines their UMR (Usual Marketing Requirements) or level of commercial imports required if the average level of imports of the past five years is to be maintained. The CSD has its own procedures for collecting the information, but no longer publishes its statistics. In the past, these appeared regularly in the FAO *Food Aid Bulletins*. But the coverage of individual donors' actions was far from complete, as many emergency and project allocations were not reported to the CSD.

The Global Information and Early Warning System on Food and Agriculture (GIEWS) was created in 1975, to monitor the world supply/

demand situation for basic foods, identify countries and regions where serious shortages were imminent, and assess possible emergency requirements. The GIEWS data are the source of information most widely used by donor agencies when assessing recipient country requirements and planning their allocations.

The GIEWS estimates and forecasts production, consumption, imports and exports, food aid needs, commitments and shipments, and cereal carryover stocks for individual developing countries. A cereals food balance is built up relating to each country's marketing year which generally corresponds to the crop calendar (the new marketing calendar begins just after the main crop harvest period). Food aid figures also relate to this calendar but they take into account undelivered pledges from the previous year. Donors are regularly asked by telex to update information on their pledges. Information from recipient countries is received through WFP and FAO representatives or directly from national early warning systems or statistical sources. Information from other sources, including news media and other agencies, is also incorporated.

The data are presented and analysed in four publications. *Food Outlook* (11 issues a year) gives an up-to-date analysis of the global food situation and an assessment of short-term production prospects, stocks, exports, imports, food aid and prices. It covers cereals and other imported food stuffs. Its *Statistical Supplement* gives a longer time series, but only for the main producing or trading countries. *Foodcrops and Shortages* (11 issues a year) describes crop conditions, production prospects and the national food supply situation (including import requirements and food aid commitments) in both developed and developing countries. It identifies countries where current crop conditions give cause for concern or which have an acute food shortage. But its circulation is restricted to governments, international organisations and some NGOs. *Special Alerts* are issued when the food supply in a specific country or a group of countries threatens to deteriorate. They are the primary source for early warnings and an evaluation of food aid requirements.

Since 1984 a special report on the *Food Supply Situation and Crop Prospects in Sub-Saharan Africa* has appeared every two or three months. It contains production, stocks, and cereal import and export information, focusing particularly on food aid requirements, pledges and deliveries. More restricted in coverage is the confidential *Sahel Weather and Crop Situation Report*, transmitted by telex between July and November (approximately ten times a year). This gives a regional summary of the food and agrometeorological situation of each CILSS[2] country during the growing season, and the final issue reports on the growing season.

World Food Programme (WFP)

WFP monitors world-wide food aid allocations, shipments and receipts in considerable detail. It has also long been the only source of information on the different uses of food aid (programme, emergency and various project uses) resulting from an annual survey by its Country Offices in recipient countries. WFP data, including INTERFAIS (discussed below), are summarised in the annual report on *Food Aid Policies and Programmes* (now published), presented for review by its governing body, the Committee on Food Aid Policies and Programmes (CFA). They are also disseminated through FAO publications such as *Food Aid in Figures*.

During the 1983–85 food crisis in Africa, a more detailed database for African countries was developed as part of an international attempt by donors to co-ordinate their pledges and shipments in response to the African emergency. This was intended to improve programming decision making and to facilitate better co-ordination of shipping and overland transport to destinations where handling and transit facilities were severely limited. The database was computerised in 1985 at the International Computer Centre (CIC) in Geneva. The reports, *Status of Food Aid Deliveries to African Countries Affected by the Food Crisis*, were circulated monthly and bi-monthly. They detailed the commitments, deliveries, nature, origin and type of use of the food aid supplied for sale, emergency, and project use on a country-by-country basis.

Since 1987, data on individual food aid actions from commitment through to delivery have been drawn together by WFP as the International System of Information on Food Aid (INTERFAIS). It is intended to centralise detailed information on all food aid which is transmitted monthly by the donors and approximately 100 WFP offices in recipient countries. INTERFAIS represents the most detailed source of information on food aid in a disaggregated form. Installed at the CIC in Geneva as well as in sub-systems in recipient countries, the databank will soon be accessible to all the members of INTERFAIS by direct electronic mail.

While setting up an INTERFAIS network during the summer of 1988, WFP looked into the reformulation of its regular reports on food aid deliveries to sub-Saharan Africa, publication of which had been suspended. During 1989, it began to circulate to INTERFAIS participating agencies a quarterly *Food Aid Monitor* which provides detailed information on schedules and deliveries for each individual recipient country and donor (including, for the first time, the NGOs) by use (project, emergency, etc.), beneficiary type (for example, refugees) and commodity type. More detailed data on delivery schedules for the operational use of food aid for countries seriously affected by emergencies is also included, as is logistical information for management co-ordination. These data provide the most detailed cross-sectional picture of food aid, but so far only for the short period since the beginning of INTERFAIS in 1987. As the data are not strictly comparable with earlier series, it will only be after a number of years that researchers can begin to use INTERFAIS data for time series analysis.

Organisation for Economic Co-operation and Development (OECD)

The Co-operation and Development Division of OECD follows the aid politics and budgets of member countries of the Development Assistance Committee (DAC) in order to assess and compare their official development assistance (ODA). As well as information on overall financial flows, their destination, terms, conditions, and purposes, etc., the OECD asks member countries for detailed information on their food aid allocations for the calendar year (January/December) in two questionnaires: one elicits the tonnage of food aid by destination and product, and the other the value, using international prices, of food aid by type, purpose and destination, including transport costs and cash contributions. The OECD is presently the only global source for value data, that is the value attributed by donors to the commodities, and the associated costs. The data on each donor member of the Organisation are only published annually in the *Development Co-operation Review*, but detailed statistics are available on request at the OECD Secretariat.

International Wheat Council (IWC) and Food Aid Committee (FAC)

The International Wheat Council was established in 1949 under the first of a series of International Wheat Agreements (IWA). The first IWAs involved regulatory supply and price provisions, but, as the structure of the world grain markets changed, so the role of the Council evolved. It is now principally a forum for discussion of world grain problems and a centre for the collection and dissemination of information, including detailed and comprehensive statistics on world grain markets.

Since 1967, each agreement has consisted of two legally distinct conventions, a Wheat Trade Convention and the Food Aid Convention (FAC). Membership of the FAC consists of most OECD countries and Argentina. Members undertake to supply certain minimum quantities of grain as food aid for the benefit of developing countries. Over the years, the FAC has been expanded in its coverage and increased its commitments. The most recent negotiation was in 1986.[3]

Detailed records are kept to monitor members' fulfilment of their commitments, and these are regularly published by the International Wheat Council on behalf of the FAC. The reporting period is July/June (which corresponds to IWC trade statistics). Food aid shipments are shown in tonnages and also converted at agreed rates into grain equivalent, which is the basis for members' obligations. These data, by definition, only cover cereals food aid, and as donors are not obliged to report allocations or shipments outside their obligations under the FAC, the reporting is incomplete as suggested by comparison with the FAO and WFP data series.

Bilateral Donor Agencies

Most donor organisations produce statistics for their own programmes as part of their internal reporting and in accounting for aid activities to legislative assemblies and auditing bodies. They also supply information to FAO, IWC, OECD and WFP. Space precludes a detailed discussion of several food aid donors, so this brief survey of sources describes only the major one, the United States.

Two services of the US Department of Agriculture (USDA) deal with world-wide statistics: the Foreign Agriculture Service, which publishes the *Foreign Agriculture Circular* (on grains) and an agricultural statistics yearbook, and the Economic Research Service (ERS), which establishes food balances using a computerised econometric model. Only this second service gives specific information on food aid which is available in the quarterly publication *World Food Needs and Availabilities*. This parallels the GIEWS *Food Crops and Shortages* reports in estimating national food aid 'needs'. The ERS methodology for forecasting needs involves either assuming there will be 'no change' in per capita consumption, or satisfying 'nutritional needs' based on FAO/WHO standards. The basic methodology and way of calculating an individual country's capacity to import commercially are also described in each report. In addition, USDA monitors the situation in some of the main importing countries, giving more details on food aid in periodic telexes or documents.

USDA and the United States Agency for International Development (USAID) regularly report on food aid activities, for example in the annual report to Congress, *Food for Peace*. These reports are an important source of historical documentation for the most significant donor. Reporting is on the basis of the United States' fiscal year (October/September) in current dollars and according to the various 'titles' of the Public Law 480 legislation under which food aid is provided. These are distinctive reporting features of the United States institutions, but naturally other donors do much the same according to their own legislative and administrative framework.

Following the drought in Africa in 1983/84, the Africa Bureau of USAID created a *Famine Early Warning System (FEWS)* to follow the food situation in the most vulnerable African countries: Burkina Faso, Chad, Ethiopia, Mali, Mauritania, Mozambique, Niger and Sudan. Since 1986, it has been publishing regular country reports in a statistical format, providing background data, assessment of current crops based primarily on remote sensing data from NASA on USDA/NOA Joint Weather Facility, and some information on food aid arrivals and distributions. Thematic issues common to several countries are also prepared.

ACTUAL SITUATION AND POSSIBLE IMPROVEMENTS

Sources of Discrepancies in Food Aid Statistics

The above account of organisations involved in generating and collating food aid statistics might lead observers to believe that the analyst's job is an easy one. In fact it can be difficult to compare these different sources of information because they do not generally have the same objectives. Some of them build historical series, others introduce food aid into food aid balances, others compare food aid allocations to commitments or to the level of commercial imports. The main characteristics of all these statistical sources are summarised in Table 1. Different rules are used to collate food aid statistics, reflecting the different priorities. Three main factors account for discrepancies between statistical data.

(a) Reporting Periods

There are four different reporting periods and some of these even differ from one country to another.
- The *calendar year* (January/December) is used by OECD for comparisons of ODA, and by WFP's INTERFAIS and FAO for noncereal food aid statistics.

TABLE 1
IMPORTANT FOOD AID STATISTICAL SOURCES: REPORTING PERIODS, COVERAGE, OBJECTIVES AND PUBLICATIONS

SOURCE OF INFORMATION	REPORTING PERIOD	COMMODITIES	MAIN OBJECTIVES	PUBLICATIONS
FAO-GIEWS	Marketing year of the country	Cereals	National food balance sheets, food situation	Food Outlook, Foodcrops and Shortages, Africa Report
FAO Food Security Group	July/June	Cereals and non-cereals	Historical series	Food Aid in Figures
C.S.D	–	Cereals	Comparison with commercial imports	–
WFP/INTERFAIS	All periods	Cereals and non-cereals	Management and logistics	Food Aid Monitor
IWC/FAC	July/June	Cereals	Comparison with food aid commitments	FAC Annual Report
OECD/DAC	Jan/Dec	All food aid	Valuation, aid historical series	OECD Development Co-operation Review
USDA	–	All	National food balance sheets	World Food Needs and Availabilities
USAID (FEWS)	–	All	Food situation in disaster-prone countries	Country Reports

SOURCES OF DATA FOR FOOD AID ANALYSIS

- The *fiscal year* is the calendar year in the case of the European Community. However, in Canada and Japan, the fiscal year begins on 1 April whereas in the United States it begins on 1st October.
- The IWC uses the split, agricultural, year (July/June) for reporting on food aid allocations within the Food Aid Convention. FAO's Food Security Group also chose this reporting period for its historical series.
- Finally, to integrate food aid into food balance calculations, it is necessary to use the *marketing year* which normally corresponds to the crop calendar which differs from country to country. Amongst low-income food-deficit countries there are eight different reporting periods, and even in the same sub-region there are often three or four different ones.

(b) Conversion Factors

To aggregate food aid figures for different commodities, conversion factors are required. Cereals aid is usually expressed in 'grain equivalent', but conversion factors may differ amongst sources (especially for rice because of the conversion factor from paddy, 'rough' rice, to milled rice) and for wheat (wheat equivalent for wheat flour, bulgur, etc.). Difficulties also arise with composite commodities or blended foods which contain cereal and non-cereal components.

(c) Date of Record

Another important factor which renders comparisons of food aid statistics very difficult, especially between the donor countries or organisations and the recipient countries, is the problem of the recording stage. Is a food aid action reported in the table when it is decided upon, allocated, shipped, received in the recipient country or distributed to final beneficiaries? As a long period can elapse between allocation and the execution of an 'action', some actions can be recorded in two consecutive years. So it is necessary to know if the action is still only a pledge or if the food has been delivered.

Possible Improvements

All these factors explain why big differences can be found between statistical sources. As an illustration, a recent study by Egg and others [*1987*] on cereal policies and cereal trade in Western Africa, funded by the Club du Sahel and the French Ministry of Co-operation and conducted by French-African research teams, made an inventory of cereal imports in the 18 countries of the region, using seven international sources of statistics. In some countries, large discrepancies appear in the tables comparing the figures. Initially, statistics from only one source, FAO's Food Security Group, were introduced into the databank, but in the second phase of the programme the addition of other sources is foreseen. Such database studies are a good point of reference for comparing statistics, analysing the differences and improving their reliability. They can be a first step to developing better co-ordination between different organisations or even services within the same organisation. Thus discussions aimed at improving food aid statistics should consider two ideas.

First, that organisations should adopt similar rules for monitoring and presenting food aid statistics. For instance, an international agreement on conversion factors could help to eliminate the differences between sources of information.

Second, they should attempt to simplify data collection. Currently, each data collecting organisation asks donors for separate food aid reports. This multiplication of questionnaires is time-consuming for donor staff who may sometimes under pressure not have the time to answer correctly and send regularly updated information to all the organisations. This in turn increases the differences between statistical sources.

A more ambitious conception would be to agree on a common international code for each food aid allocation which could be used by all statistical agencies. This code would follow each allocation from pledge to delivery, thus eliminating double counting. It would also facilitate comparison between statistical sources which use different reporting periods. Donor co-operation in creating the management and logistics database for Africa in 1984, and the subsequent expansion of this management information tool into the global INTERFAIS stored at the CIC after 1987, indicate at last a recognition on the part of donor agencies of the benefits of moving in that direction from a purely operational standpoint.

Finally, a discussion among agencies involved in food aid statistics could lead to a sharing of responsibilities to avoid duplication. Each organisation could concentrate on its specific area of interest: valuation of aid flows, record of pledges, logistics, etc.

These proposals would have to be discussed between all the organisations involved. FAO or WFP could play a central role in organising collaboration and in proposing international definitions and rules for food aid statistics. As these organisations react to their members' wishes, some food aid donors or recipient countries could put these proposals at a future session. Discussion of them among the agencies involved in food aid statistics could be a first step towards an international code on food aid which would help analysts, especially decision makers, to define food aid projects and food security policies.

NOTES

1. See also Chapter 6 where Clay and Benson discuss these data for produce acquired in developing countries.
2. 'Comité Inter-Etats de Lutte contre la Sécheresse au Sahel': Burkina Faso, Cape Verde, Chad, Gambia, Guinea Bissau, Mali, Mauritania, Niger and Senegal.
3. See Parotte [*1983*] for earlier history of the FAC.

REFERENCES

Egg, J., J.-J. Gabes, and J.P. Lemelle, 1987, *From a Single Regional Market to Regional Markets*, Paris: Institut de Recherches et d'Applications des Méthodes de Développement; Montpellier: Institut National de la Recherche Agronomique; and Cotonou: Université Nationale du Bénin.

Parotte, J.H., 1983, 'The Food Aid Convention: Its History and Scope', *IDS Bulletin*, Vol.14, No.2, University of Sussex.

Index

access 119, 130–34, 186
additionality 6, 7, 12, 20, 33, 37–65, 91–2, 94
Africa
  Sub-Saharan 8, 10, 11, 12, 18, 19–26, 29, 30, 31, 32, 67, 70–71, 91–115, 143–78, 180–90, 191–201
  West 177, 198
African food crisis 6, 9, 16, 19–26, 32, 33, 150–51, 194
agricultural production 4, 17, 100–101, 103, 108, 118, 181, 187
  impact of food aid on 15–18, 29, 32, 33, 39, 55, 66–89, 96, 187, 189
appraisal 122–5, 146
Argentina 40, 58–64, 145, 152, 153, 161, 196
Asia 19, 154, 171
  South 32
Australia 8, 27, 42, 58–64, 149, 171, 173, 177
Austria 173

balance of payments support 4, 16, 30, 37, 50–51, 52–7, 68, 84, 94–5, 98, 110, 119, 128, 182, 183–4, 187–8
Bangladesh 8, 18, 25, 26, 28, 29, 33, 70, 145, 160
Belgium 173
Benin 158, 165
Bhutan 160
bilateral food aid 8, 39, 92, 97, 105, 166, 169, 170, 192, 196–7
  see also names of individual donor countries
Bolivia 161
Botswana 20, 25, 53, 71, 72, 91–115, 148, 158, 174
bovines *see* cattle
budgetary support 37, 50–51, 68, 85, 94–5, 182
buffaloes *see* cattle
Burkina Faso 53, 158, 197, 201
Burma 148, 152, 153, 154, 160, 166
Burundi 157, 158

butter 40
  oil 40, 71, 129
  see also ghee

Cameroun 71, 158, 165
Canada 2, 6, 27, 40, 41, 42, 58–64, 173, 199
Cape Verde 162, 201
CARE 22
cash crops 87, 163
Catholic Relief Services (CRS) 22
cattle 105, 117–41
CCC *see* Commodity Credit Corporation
cereals
  as food aid 2, 8, 9–10, 12–14, 21, 26, 28, 31, 32, 40, 47, 53–7, 67, 70, 71, 79, 82, 102, 107, 108–9, 152, 154, 164, 166, 170, 173, 177, 181, 192, 193, 196, 198, 199
  exports 162, 163, 166
  imports 18, 20, 28, 51–2, 53–7, 67, 79, 87, 108–9, 188, 199
  markets 7, 25, 27, 28, 39, 48, 75, 79–81, 83–4, 86, 176, 177, 199–200
  production 103, 108–9, 151, 154, 163, 193
  see also commodity surpluses
CFA *see* Committee on Food Aid Policies & Programmes
Chad 20, 158, 197
China 152, 153, 160, 201
CIC *see* International Computer Centre
CILSS *see* Comité permanent inter-états de Lutte contre la Secheresse dans le Sahel
Club du Sahel 199–200
coarse grains 26, 32, 40, 52, 145, 148, 154, 155, 156, 170, 171
  see also maize; millet; sorghum
Colombia 17, 86, 161
Comité permanent inter-états de Lutte contre la Secheresse dans le Sahel (CILSS) 146, 193–4, 201
Committee on Food Aid Policies &

202

INDEX

Programmes (CFA) 148, 194
Committee on Surplus Disposal (of the FAO) 2, 52, 192
  see also Usual Marketing Requirements (UMR)
commodities see names of individual food aid commodities
commodity
  acquisition 26, 32, 167–8
  exchanges 26, 143–5, 151, 154, 156–7, 168, 171–2, 174
  surpluses 1–2, 6, 7, 9–11, 12, 13, 27, 37, 39, 40–41, 43, 48–9, 106, 118, 128–9, 157, 163, 172–4, 190
  see also Committee on Surplus Disposal; exports; triangular transactions
Commodity Credit Corporation (CCC) 51
Common Agricultural Policy see European Community, Common Agricultural Policy
Compensatory Financing Facility (of the IMF) 29, 188
concessional sales 18, 33, 46, 47, 50, 52–7, 92
co-operatives, dairy see dairy co-operatives
Costa Rica 161
Côte d'Ivoire 158, 165
counterpart funds 56, 84, 95, 182, 185, 189
  see also food aid monetisation
cows see cattle
CRS see Catholic Relief Service

DAC see Development Assistance Committee (of the OECD)
dairy aid 7, 8, 11, 13–14, 22, 27, 38, 40, 42, 48, 52, 67, 87, 117–41
  see also commodity surpluses
dairy co-operatives 117–41
dairy development 4, 15, 29, 33, 117–41
  see also Operation Flood
data on food aid see statistics on food aid
delivery systems see transportation
Denmark 166, 173
Development Assistance Committee (of the OECD) 1, 8, 42, 50, 195, 198
diamond investment 103, 111
disincentive effect see agricultural production, impact of food aid on
displaced populations 3–4, 26, 176
  see also refugees

domestic trade 17, 39, 53, 71, 97, 168
donor policy 1–2, 3, 4, 5–6, 18, 19, 20, 22, 23, 25, 26, 27–8, 38–49, 56, 69, 84, 85, 105–6, 118, 128, 138, 143, 148, 150–51, 163–6, 171, 174, 175–7, 181, 184, 192, 193, 194, 196, 197–200
donors see names of individual countries
dried skimmed milk see skimmed milk powder
drought 16, 19–20, 25, 27, 28, 34, 48, 82, 98, 99, 105–6, 107, 110–113, 150, 170, 197

early warning systems 20, 23, 72, 75–8, 79–86, 98, 193
  see also INTERFAIS; Global Information & Early Warning System
EC see European Community
econometric models 17, 18, 54–5, 58–64, 69–70, 82, 91–115
edible oils 2, 7, 8, 14, 27, 40, 41, 149
Egypt 18, 53
El Salvador 161
emergency aid 3–4, 11, 13, 21, 23–5, 26, 29, 30, 31, 33, 39, 55–6, 71, 79–82, 146, 156–7, 167, 168, 172, 175–6, 181, 189–90, 192, 194, 195
employment see labour
Equador 161
equity 130–38, 140, 184, 186
Ethiopia 20, 22, 23, 34, 66, 71, 76, 78, 79–81, 86, 87, 158, 169, 176, 197
Euronaid 22
Europe 149, 170
European Community (EC) 2, 11, 15, 18, 22, 27, 33, 39, 40, 42, 53, 58–64, 93, 97, 105, 106, 117, 118, 128–9, 138–9, 144, 146, 148, 162, 166, 170, 173, 178, 188, 199
  Common Agricultural Policy 15, 39
evaluation 19, 69, 87, 92–3, 95, 119, 146–7, 193
exports 9–11, 18, 44, 54–7, 58–64, 100–101, 118, 128, 145–6, 149–50, 151, 157, 162, 163, 166, 167, 169, 174, 183, 185, 190, 193

FAC see Food Aid Convention
FAO see Food and Agriculture Organisation
Fiji 160
financial aid 7, 31, 37, 38, 41–3, 49–50,

203

52, 57, 68, 168, 169, 176, 182, 183, 184, 188, 189
fish products 2, 8, 40, 148, 156
food
  consumption 17, 18, 50, 52–3, 57, 66–89, 94–5, 110–11, 140, 143, 167, 172, 193, 196
  prices 6, 11, 16, 26, 40, 43–6, 50–52, 58–64, 66–89, 98, 99, 168, 187, 193, 195
  security 13, 16, 18, 25, 26, 27–9, 76, 87, 91–115, 118, 163, 169, 170, 172, 174–7, 182, 183, 188, 201
  strategy 4, 18
food aid
  cost 7–8, 43–8, 58–64, 94, 106, 168, 184
  cost-effectiveness 12–13, 31, 37–57, 96, 143, 146–7, 151, 167–72, 177, 186–7
  monetisation 5, 145, 185
  value 7–8, 50–52, 94, 195
  see also statistics on food aid
Food Aid Convention (FAC) 2, 8, 9, 12, 27, 28, 33, 34, 39, 40, 144, 145, 147, 166, 192, 195–6, 198, 199, 201
Food and Agriculture Organisation (FAO) 33, 40, 54–6, 57, 92, 93, 97, 106, 110, 146–7, 156, 192–3, 194, 196, 197, 198, 200
  see also Committee on Surplus Disposal; Global Information & Early Warning System
food production *see* agricultural production
food-for-work 4, 5, 25–6, 32, 33, 34, 67, 69, 71–2, 75, 79, 95, 106, 185
foreign aid 2, 3, 4, 6, 138–9
France 39
France, Ministry of Co-operation 199–200

Gambia 158, 201
Germany 40, 166, 173
Ghana 158, 165
ghee 133, 132, 137
Global Information & Early Warning System (GIEWS) 30, 192–3, 196, 198
Green Revolution 15, 87, 125, 133
Guatemala 161
Guinea Bissau 158, 201
Guyana 161

Honduras 161
human capital 4, 25, 33, 185–6

IBRD *see* World Bank
IEFR *see* International Emergency Food Reserve
IFPRI *see* International Food Policy Research Institute
ILO *see* International Labour Organisation
IMF *see* International Monetary Fund
imports 8, 11, 18, 20, 28, 29, 32, 51–2, 53–7, 67, 79, 84, 86, 87, 88, 91–115, 125, 167, 172, 181, 183, 184, 192, 193, 196
income distribution 15, 32, 82, 99, 130–34, 184–5
India 15, 17, 18, 28, 29, 33, 53, 67, 69, 70, 86, 117–41, 145, 157, 160, 187
Indian Dairy Corporation *see* National Dairy Development Board, India
Indonesia 160
information on food aid *see* statistics on food aid
Institute of Social Studies, the Hague 15
Integrated Rural Development Programme (IRDP) 133
INTERFAIS (of the WFP) 30, 33, 194–5, 197, 198, 200
International Bank for Reconstruction and Development *see* World Bank
International Committee of the Red Cross 173, 192
International Computer Centre (CIC) 194, 200
International Emergency Food Reserve (IEFR) 27
International Food Policy Research Institute (IFPRI) 181
international food trade *see* world food markets
International Labour Organisation (ILO) 107
International Monetary Fund (IMF) 29, 97, 98, 99
  see also Compensatory Financing Facility
International System of Information on Food Aid *see* INTERFAIS (of the WFP)
International Wheat Council 46, 51, 192, 195–6, 198, 199
IRDP *see* Integrated Rural Development Programme
ISS *see* Institute of Social Studies, the Hague
Italy 173
Ivory Coast *see* Côte d'Ivoire
Japan 2, 3, 33, 40, 41, 42, 46, 154, 166,

INDEX

173, 178, 199

Kampuchea 148–9
  see also Thai–Kampuchean Border Relief Operation
Kenya 20, 33, 69, 71, 151, 154, 158, 162, 165, 169, 175
khoa 131, 137
Korea, Republic of *see* South Korea

labour 26, 66–89, 105, 130, 134, 139, 182, 184–5
land 132–3
  landless 130–31, 132, 133
  reform 133, 188–9
Latin America 178
League of Red Cross and Red Crescent Societies 192
Lebanon 159
Lesotho 25, 53, 71, 87, 107, 158
local purchases *see* triangular transactions

Madagascar 158, 188
maize 9, 108–9, 148, 149–50, 162, 169, 170, 171, 174, 175, 177, 178
Malawi 151, 154, 157, 158, 162, 165, 174, 177
Mali 20, 159, 165, 175, 197, 201
malnutrition *see* nutrition
Mauritania 159, 197, 201
Mexico 161
milk supply & production *see* dairy development; dairy aid
millet 77, 83, 84, 87, 178
modelling *see* econometric models
Mozambique 20, 146, 149, 170, 174, 176, 197
multilateral food aid 39, 49, 92, 97, 105, 111, 186, 189–90, 192

National Aeronautics & Space Administration (NASA) 197
National Co-operative Dairy Federation of India (NCDF) 120
National Dairy Development Board, India (NDDB) 117, 118–19, 120, 121, 122, 128, 129, 130, 139, 140
Nepal 157, 160
Netherlands 3, 166, 173
New Zealand 128
NGOs *see* Non-Governmental Organisations
Niger 151, 159, 165, 174, 197, 201
Non-Governmental Organisations (NGOs) 3, 19, 20, 22–3, 33, 144, 173, 177, 186, 191, 193, 195
non-project aid *see* programme aid
Norway 2, 40, 166, 173
nutrition 118–19, 139, 140, 171, 190, 196
  malnutrition 31, 82–4, 98, 185–6
  projects 4, 15, 25, 32, 76, 84, 95, 105–7, 110, 112

ODA *see* Official Development Assistance
OECD *see* Organisation for Economic Co-operation & Development
Official Development Assistance (ODA) 6, 195, 197
oil prices 19, 148
Operation Flood 15, 117–41
opportunity cost *see* food aid cost
Organisation for Economic Co-operation & Development (OECD) 20, 33, 195, 196, 197, 198
  see also Development Assistance Committee

Pakistan 145, 152, 153, 160, 166
Philippines 160
Pisani, E. 4
PL480 *see* Public Law 480
programme aid 13, 21, 23, 24, 31, 55–6, 71, 75, 157, 183, 184, 187, 194
project aid 4, 11, 13, 21, 31, 33, 53, 55–6, 95, 156–7, 174, 184, 192, 194, 195, 201
Public Law 480 6, 7, 9, 11, 13, 18, 22, 27, 38, 41, 51, 53, 67, 68, 93, 95, 197
pulses 79, 156, 162, 170, 178

rapid rural appraisal 78, 87
Red Crescent *see* League of Red Cross and Red Crescent Societies
Red Cross *see* International Committee of the Red Cross; League of Red Cross and Red Crescent Societies
refugees 3–4, 148–9, 176, 195
  see also displaced populations
rice 9, 26, 34, 40, 41, 52, 79, 83, 84, 145, 148, 149, 154, 155, 156, 166, 177, 188, 199
Rwanda 71, 157

SADCC *see* Southern African Development Co-ordination Conference
Sahel region 19, 20, 22, 150–51, 175, 176, 178, 193

205

St. Christopher & Nevis 161
St. Lucia 161
São Tomé 162
Saudi Arabia 159
Schultz, T.W. 51, 66, 73, 94
Senegal 66, 69, 77, 79, 82–4, 86, 159, 165, 175, 201
shipping costs *see* transportation costs
skimmed milk powder 13, 14, 40, 71, 128, 129
Somalia 71, 159, 171, 176
sorghum 79–82, 85, 108–9, 156, 169, 178
South Africa 98, 105, 107
South Korea 33, 53, 96
Southern African Customs Union 107, 115
Southern African Development Co-ordination Conference (SADCC) 106, 146, 176, 177
Sri Lanka 18, 53
statistics on food aid 19, 21, 30, 31, 42, 62, 70, 77, 78, 82, 85, 91–5, 99–102, 108–9, 111, 114, 146–7, 152, 153, 155–6, 158–63, 165, 168, 173, 175, 180–82, 191–201
  see also early warning systems; econometric models; INTERFAIS; Global Information & Early Warning System
structural adjustment 30, 98, 180–90
Sub-Saharan Africa *see* Africa, Sub-Saharan
Sudan 20, 22, 23, 66, 69, 71, 79, 84–5, 86, 146, 156, 159, 165, 169, 171, 176, 177, 197
supplementary feeding programmes 71, 95, 106, 107, 110
Surinam 161
surplus disposal *see* Committee on Surplus Disposal; commodity surpluses
Swaziland 107, 159, 165
Sweden 40
Switzerland 166, 173

Tanzania 33, 71, 159, 162, 165, 175
targeted food aid 5, 25, 28, 39, 76, 84, 85, 110, 119, 184, 185–6, 188
Thai–Kampuchean Border Relief Operation 148–9
Thailand 145, 148–9, 152, 153, 154, 156, 157, 160, 163, 166, 169
Togo 159, 162, 165
transportation 6, 22, 145, 149–50, 194
  costs 8, 11, 33, 46–7, 51, 56, 167–70, 174, 184, 195

triangular transactions 3, 8, 26, 143–78, 181, 183, 184, 192
trilateral exchanges *see* triangular transactions
tubers 26, 32
Tunisia 53, 70
Turkey 153, 159

Uganda 159, 165
UK *see* United Kingdom
UMR *see* Usual Marketing Requirements
United Kingdom 3, 39, 49, 166, 173
United Nations Border Relief Operations (UNBRO) *see* Thai–Kampuchean Border Relief Operation
United Nations Children's Fund (UNICEF) 185–6, 192
United Nations Conference on Trade & Development (UNCTAD) 97
United Nations Development Programme (UNDP) 97, 189
United Nations Food and Agricultural Organisation *see* Food and Agriculture Organisation (FAO)
United Nations High Commission for Refugees (UNHCR) 173, 192
United Nations Relief & Works Agency (UNRWO) 192
United Nations World Food Programme *see* World Food Programme (WFP)
United States 1, 3, 6, 7, 9, 11, 13, 18, 21, 27, 33, 37, 38, 40, 41, 42, 46, 51, 58–64, 71, 105, 146, 149, 166, 170, 171, 173, 177, 196–7, 199
United States Agency for International Development (USAID) 68, 196–7, 198
  Famine Early Warning System 197, 198
United States, Department of Agriculture 196, 198
  see also Commodity Credit Corporation
Upper Volta *see* Burkina Faso
USA *see* United States
USAID *see* United States Agency for International Development
Usual Marketing Requirements (UMR) 2, 52–3, 91, 183, 192
  see also Committee on Surplus Disposal

vegetable oils *see* edible oils

INDEX

Vietnam 160

WFC *see* World Food Council
WFP *see* World Food Programme
wheat 9, 26, 27, 33, 34, 40, 44–7, 51–2, 54, 58–64, 67, 71, 79, 83, 85, 88, 144, 147, 153, 154, 155, 177, 195–6, 199
  bread 85, 86
  flour 33, 83, 85, 145, 155, 156, 199
WHO *see* World Health Organisation
World Bank 33, 97, 98, 99, 111, 122–3, 176, 186, 188, 189
World Food Conference (1974) 28, 148
World Food Council (WFC) 4
World food crisis (1972–74) 9, 19
World food markets 7, 9, 16, 26, 27, 29, 37, 39, 40, 43–6, 50–52, 56, 58–64, 83, 94–5, 97, 98, 99, 145, 162–3, 170, 177, 195–6
  see also food prices
World Food Programme (WFP) 25, 33, 39, 93, 97, 105, 106, 111, 145, 146–7, 148, 149, 150, 152–6, 157, 162, 169–70, 173–4, 177, 186, 188, 189, 191, 192, 193–4, 195, 197, 198, 199
  see also INTERFAIS
World Health Organisation (WHO) 196

Zaire 159
Zambia 148, 159
Zimbabwe 20, 146, 148, 149–50, 151, 154, 157, 159, 162, 163, 164, 165, 166, 170, 177

Notes on Contributors

Charlotte Benson was a Research Officer at the Relief and Development Institute, London, and is an economist with the Commodities Research Unit, London.

John Cathie is a Senior Research Officer in the Department of Land Economy, Cambridge University and a Fellow and Tutor at Wolfson College, Cambridge.

Edward Clay is a Research Fellow of the Overseas Development Institute, London.

Martin Doornbos is Professor of Political Science at the Institute of Social Studies, The Hague.

Liana Gertsch is a Research Associate at the Institute of Social Studies, The Hague.

Stéphane Jost is currently with FAO's Global Information and Early Warning System. Previously, as consultant to the Club du Sahel of the OECD, he prepared annual reports on the food aid situation in Sahelian countries.

Panos Konandreas is a Senior Economist in FAO's Commodities and Trade Division.

Simon Maxwell is a Fellow of the Institute of Development Studies at the University of Sussex.

Ram Saran was formerly head of FAO's Food Security and Food Aid Policies Group.

Hans Singer is Emeritus Professor and a Fellow of the Institute of Development Studies at the University of Sussex.

Olav Stokke is a senior Research Fellow at the Norwegian Institute of International Affairs.

Piet Terhal is an Associate Professor of Development Economics and Director of the Centre for Development Planning at Erasmus University.

86373
M/L